Flood Management in the 21st century

Their Royal Highnesses
Prince John Charles Wright

Prince Joe Duncan Wright

Copyright ©HRH Princess Dillys Wright 2002

First Published in 1990

Table of Contents

Introduction 3
Flood Management 6
Improved Drainage Infrastructure **9**
Green Spaces and Permeable Surfaces 14
Sustainable Urban Planning 40
Smart Drainage Systems 54
Flood Control Structures 64
Flood Warning and Management Systems 67
Public Education and Awareness 70
Green Infrastructure Retrofitting 89
Desalination and Water Recycling 103
Infrastructure Resilience Planning 110
Collaboration and Partnerships 133

Introduction

Flood management have become increasingly important in the 21st century. Rapid urbanization and population growth have led to increased land development and expansion of impervious surfaces such as roads, buildings, and parking lots. This has reduced natural infiltration and increased surface runoff, exacerbating flooding in urban areas. Urbanization leads to extensive land development and impervious surfaces, reducing natural drainage and increasing the risk of urban flooding. Effective water drainage systems are essential to protect growing urban populations and infrastructure from flood-related damages.

Climate change is causing more frequent and intense rainfall events, leading to heightened flood risks in many regions. Rising global temperatures are also contributing to the melting of glaciers and polar ice caps, leading to sea-level rise and increased coastal flooding. Climate change has resulted in more frequent and intense precipitation events, exacerbating flood risks worldwide. Rising global temperatures contribute to the melting of glaciers and ice caps, leading to sea-level rise and increased coastal flooding. Adequate flood management strategies are crucial to adapt to these changing climate patterns and mitigate the impacts of extreme weather events.

Floods can cause severe damage to critical infrastructure such as roads, bridges, airports, and utilities. Proper flood management helps safeguard infrastructure assets, ensuring the continued operation of essential services like transportation, energy, and water supply. Protecting infrastructure from flood-related damages is vital for maintaining economic productivity and societal functioning. Aging infrastructure, inadequate drainage systems, and poor land-use planning exacerbate flood risks in many communities. Urban areas are particularly vulnerable due to the concentration of critical infrastructure and densely populated areas, increasing the potential for flood-related damages and disruptions. Deforestation, wetland destruction, and habitat loss have reduced the natural capacity of ecosystems to absorb and mitigate floodwaters. Degraded natural habitats and impaired waterways diminish the resilience of communities to flooding and exacerbate the impacts of extreme weather events.

Flooding poses significant social and economic risks, including property damage, loss of life, displacement of communities, disruption of essential services, and economic losses. Flooding results in significant economic losses due to property damage, business interruption, agricultural losses, and infrastructure repair costs. The economic impact of floods can be devastating for communities, particularly in developing countries with limited resources for recovery and reconstruction. Effective flood management measures help mitigate these economic impacts and promote long-term resilience and stability. Vulnerable populations, such as low-income communities and marginalized groups, are often disproportionately affected by flooding, exacerbating existing social inequalities. Floods can pose serious health and safety risks, including waterborne diseases, contamination of drinking water supplies, structural damage to buildings, and disruption of transportation networks. Inadequate flood management measures can jeopardize public health and safety, leading to increased morbidity and mortality rates.

Effective water drainage and flood management are essential for safeguarding water security and ensuring sustainable water resource management. Floods pose serious risks to public health and safety, including drowning, injuries, and the spread of waterborne diseases. Contaminated floodwaters can carry pathogens, chemicals, and debris, endangering human health and causing long-term health consequences. Proper water drainage and flood management are essential to protect public health, prevent waterborne illnesses, and ensure a safe living environment for communities.

Flooding can have significant environmental impacts, including habitat destruction, soil erosion, water pollution, and loss of biodiversity.

Effective flood management practices help minimize these environmental impacts by preserving natural habitats, protecting water quality, and maintaining ecosystem resilience. Sustainable flood management strategies promote the conservation and restoration of natural ecosystems, contributing to overall environmental health and biodiversity conservation. Building resilience to floods is essential for communities to withstand and recover from disaster events. Effective flood management measures, such as early warning systems, floodplain zoning, and community preparedness initiatives, help communities adapt to flood risks and minimize the social, economic, and environmental impacts of flooding. Strengthening community resilience enhances the ability of individuals and communities to cope with and recover from flood-related disasters. – HRH Prince John Charles Wright

Flood Management

Heavy rainfall events that can lead to flooding due to limited natural drainage systems. Here are some solutions that can be implemented to prevent flooding during heavy rainfall.

1. Improved Drainage Infrastructure

- Invest in upgrading and expanding the city's drainage network, including stormwater drains, culverts, and retention ponds, to enhance capacity and efficiency in managing heavy rainfall runoff.

2. Green Spaces and Permeable Surfaces

- Increase the presence of green spaces, parks, and permeable surfaces within the city to absorb and infiltrate rainfall, reducing surface runoff and alleviating pressure on drainage systems.

3. Sustainable Urban Planning

- Integrate sustainable urban planning practices, such as low-impact development (LID) techniques, green infrastructure, and water-sensitive urban design, into new developments and redevelopment projects to mitigate the impact of heavy rainfall on drainage systems.

4. Smart Drainage Systems

- Implement smart drainage systems equipped with sensors, telemetry, and real-time monitoring capabilities to optimize drainage operations, detect blockages, and manage stormwater more effectively during heavy rainfall events.

5. Flood control structures

Construct flood control structures allowing for proactive measures to be taken to mitigate damage and ensure public safety.

6. Flood Warning and Management Systems

- Develop and deploy flood warning systems to provide timely alerts to residents and authorities about impending heavy rainfall and potential flooding risks, allowing for proactive measures to be taken to mitigate damage and ensure public safety.

7. Public Education and Awareness

- Educate residents, businesses, and visitors about the risks of flooding during heavy rainfall and the importance of responsible stormwater management practices, such as avoiding littering and maintaining clear drainage channels.

8. Green Infrastructure Retrofitting

- Retrofit existing infrastructure with green infrastructure features, such as green roofs, rain gardens, and bioswales, to increase infiltration capacity and reduce runoff in built-up areas.

9. Desalination and Water Recycling

- Explore options for increasing water supply resilience through desalination and water recycling initiatives, reducing reliance on rainfall-dependent sources and minimizing the impact of heavy rainfall variability on water availability.

10. Infrastructure Resilience Planning

- Incorporate climate change adaptation measures into infrastructure planning and design to enhance resilience to extreme weather events, including heavy rainfall and flooding.

11. Collaboration and Partnerships

- Foster collaboration and partnerships among government agencies, private sector stakeholders, academic institutions, and communities to develop and implement holistic solutions for managing heavy rainfall and reducing flood risk.

Implementing these solutions and adopting a comprehensive, multi-disciplinary approach, cities can enhance their resilience to heavy rainfall events and minimize the risk of flooding, ensuring the continued sustainability and livability of the city.

Improved Drainage Infrastructure

Plan:

- Invest in upgrading and expanding the city's drainage network, including stormwater drains, culverts, and retention ponds, to increase capacity and efficiency in managing heavy rainfall runoff. In coastal areas where flooding can occur due to heavy rainfall or storm surges, implementing robust drainage infrastructure is crucial for mitigating flood risks.

Action

1. Pump Stations

Install pump stations equipped with high-capacity pumps to efficiently remove excess water from low-lying areas and prevent inundation during heavy rainfall or storm surges. Pump stations are effective drainage systems because they actively move water from low-lying or flooded areas to higher ground or treatment facilities, preventing flooding and water damage. They can be designed to handle varying volumes of water and are versatile enough to be used in various settings, including urban areas, industrial sites, and agricultural fields. Additionally, pump stations can be equipped with sensors and automation technology for efficient operation and maintenance.

2. Stormwater Drains

Construct an extensive network of stormwater drains with sufficient capacity to collect and convey rainwater away from urban areas and towards retention ponds or discharge points.

Stormwater drains are effective drainage systems because they collect and channel rainwater away from roads, buildings, and other structures, preventing flooding and water damage. They are typically integrated into the urban infrastructure and are designed to handle large volumes of water quickly and efficiently. Stormwater drains also help to reduce erosion, improve water quality by capturing pollutants, and protect the environment by preventing runoff from carrying contaminants into natural water bodies. Additionally, they are relatively low-cost compared to other drainage solutions and require minimal maintenance.

3. Canals and Channels

Develop canals and channels along coastal areas to channel excess water away from populated areas and into designated drainage basins or the sea. Canals and channels serve as effective drainage systems because they provide a natural and efficient way to redirect water flow away from areas prone to flooding or waterlogging. They can be engineered to accommodate large volumes of water and can cover extensive areas, making them suitable for both urban and rural environments. Canals and channels also help in irrigation, transportation, and water management, making them multifunctional infrastructure assets. Moreover, they can be designed with features such as locks, gates, and weirs to regulate water levels and prevent overflow during heavy rain events.

4. Retention Ponds

Build retention ponds or detention basins strategically throughout the city to temporarily store stormwater runoff during peak rainfall events, reducing the risk of flooding downstream. Retention ponds are effective drainage systems because they provide a natural and environmentally friendly way to manage stormwater runoff. They work by collecting rainwater and allowing it to gradually infiltrate into the ground or evaporate, reducing the volume of water entering the drainage system and mitigating the risk of flooding downstream. Retention ponds also help to improve water quality by trapping sediment and pollutants before they reach natural water bodies.

Retention ponds can enhance biodiversity by creating habitats for various plant and animal species. They are cost-effective to build and maintain compared to traditional drainage infrastructure and can be integrated into green spaces, providing aesthetic and recreational benefits to communities. Overall, retention ponds offer a sustainable solution to stormwater management while promoting ecological resilience.

5. Green Infrastructure

Incorporate green infrastructure features such as bioswales, vegetated swales, and rain gardens into urban design to absorb and filter stormwater runoff naturally, reducing the burden on conventional drainage systems. Green infrastructure refers to natural or nature-based solutions for managing stormwater and reducing the impacts of urbanization on the water cycle. It includes features such as green roofs, permeable pavements, rain gardens, bioswales, and urban forests. Green infrastructure is considered one of the best drainage systems for several reasons:

A. Stormwater Management
Green infrastructure mimics natural hydrological processes by allowing rainwater to infiltrate into the soil, evaporate, or be

taken up by vegetation, thus reducing runoff and the burden on traditional drainage systems.

B. Water Quality Improvement
These systems filter and treat stormwater, removing pollutants and contaminants before they enter rivers, lakes, and other water bodies, thus improving water quality and protecting aquatic ecosystems.

C. Flood Mitigation

By reducing runoff and increasing infiltration, green infrastructure helps to mitigate the risk of flooding and water damage in urban areas, particularly during heavy rain events.

D. Urban Heat Island Mitigation
Green infrastructure features such as green roofs and urban forests help to lower temperatures in urban areas, mitigating the urban heat island effect and improving overall comfort and livability.

E. Biodiversity and Habitat Creation
Green infrastructure supports biodiversity by providing habitats for various plant and animal species, contributing to urban ecological resilience and enhancing urban biodiversity.

F. Aesthetic and Recreational Benefits
Green infrastructure elements such as parks, green spaces, and urban forests provide aesthetic value and recreational

opportunities for residents, improving the quality of life in urban areas.

Green infrastructure offers a sustainable and holistic approach to drainage and stormwater management, integrating ecological, social, and economic benefits into urban planning and development.

Green Spaces and Permeable Surfaces

Plan:

- Integrate green infrastructure elements such as permeable pavements, green roofs, rain gardens, and bioswales into urban planning to absorb and infiltrate rainwater, reducing runoff and alleviating pressure on drainage systems.

Action

In flooding-prone areas with heavy rainfall, permeable pavements can be strategically employed to manage stormwater and mitigate flooding.

1. Site Assessment

- Conduct a thorough assessment of the area's topography, soil characteristics, groundwater levels, and existing drainage infrastructure to identify flooding hotspots and areas where permeable pavements can be most effective.

- Conduct a thorough assessment of the area to identify flood-prone zones, drainage patterns, and soil conditions.

- Determine the existing infrastructure and potential locations for permeable pavement installation.

Identifying Vulnerable Areas. Risk assessment involves identifying areas prone to flooding or with poor drainage. By mapping these vulnerable areas, urban planners and engineers can prioritize where permeable pavements should be installed to maximize their effectiveness.

A. Assessing Permeability Potential

Not all areas are suitable for permeable pavements. Risk assessment involves evaluating the permeability potential of the soil in different locations to determine where permeable pavements can be most effective. This ensures that the pavements will allow water to infiltrate into the ground rather than contributing to surface runoff.

B. Analyzing Design and Installation Risks

Risk assessment evaluates the design and installation process of permeable pavements. Potential risks such as improper installation, inadequate maintenance, or structural failures are identified and mitigated through proper design, construction standards, and ongoing maintenance protocols.

C. Quantifying Flood Reduction Benefits

Risk assessment helps quantify the potential benefits of permeable pavements in reducing flood risks. This involves analyzing factors such as the volume of water that can be stored within the pavement structure, the rate of infiltration into the ground, and the overall impact on reducing surface runoff during rainfall events.

D. Monitoring and Adaptation

Risk assessment involves setting up monitoring systems to track the performance of permeable pavements over time. By continuously monitoring factors such as infiltration rates, water quality, and pavement condition, authorities can identify any issues early on and implement necessary adaptations to ensure the continued effectiveness of the pavements in mitigating flooding risks.

E. Public Education and Engagement

Risk assessment also considers the importance of public education and engagement in flood risk mitigation efforts. By raising awareness about the benefits of permeable pavements and providing guidance on their proper use and maintenance, communities can ensure widespread adoption and maximize the effectiveness of these flood mitigation measures.

2. Strategic Placement

- Target areas prone to flooding or where runoff accumulates, such as parking lots, driveways, walkways, and low-lying sections of roads.

- Prioritize locations where traditional impervious surfaces exacerbate runoff and contribute to flooding.

A. Targeting Vulnerable Areas

Strategic placement involves identifying and targeting areas that are particularly vulnerable to flooding. This could include low-lying areas, areas with poor drainage systems, or locations prone to flash flooding. By focusing on these vulnerable areas, permeable pavements can help intercept and manage stormwater before it contributes to flooding.

B. Reducing Surface Runoff

Permeable pavements installed strategically in areas with high impervious surface coverage, such as parking lots, driveways, and sidewalks, can significantly reduce surface runoff during rainfall events. Instead of water quickly flowing over impermeable surfaces and inundating nearby areas, permeable pavements allow water to infiltrate into the ground, thus reducing the volume and velocity of runoff.

C. Promoting Infiltration

Placing permeable pavements in locations with suitable soil conditions can maximize infiltration rates. Areas with compacted or poorly draining soils can benefit greatly from permeable pavements, as they allow water to percolate into the ground more effectively. This helps recharge groundwater resources and reduces the risk of surface runoff overwhelming drainage systems.

D. Integration with Green Infrastructure

Strategic placement often involves integrating permeable pavements with other green infrastructure practices, such as rain gardens, bioswales, and vegetated swales. This holistic approach to stormwater management helps create a network of interconnected systems that work together to capture, treat, and infiltrate stormwater runoff, thus reducing the overall impact of flooding.

E. Protecting Critical Infrastructure

Permeable pavements can be strategically placed around critical infrastructure such as hospitals, schools, and emergency response facilities to protect them from flooding. By intercepting stormwater runoff before it reaches these facilities, permeable pavements help maintain access and functionality during extreme weather events, ensuring the safety and well-being of the community.

F. Community Engagement and Planning

Strategic placement involves engaging stakeholders and community members in the planning process to identify priorities and concerns related to flooding. By incorporating local knowledge and feedback, planners can ensure that permeable pavements are placed in locations where they will have the most significant impact on reducing flood risk and improving community resilience.

3. Pavement Selection

- Choose appropriate permeable pavement materials based on site-specific factors, traffic loads, and design requirements.

- Options include porous asphalt, pervious concrete, permeable interlocking concrete pavers, and permeable gravel.

Pavement selection can significantly contribute to mitigating flooding by influencing the amount of stormwater runoff generated and how it is managed. Here's how:

A. Permeability

Choosing permeable pavements, such as porous asphalt, pervious concrete, or permeable interlocking concrete pavers, allows rainwater to infiltrate into the ground, reducing surface runoff. Unlike traditional impermeable pavements, which shed water quickly, permeable pavements help recharge groundwater, alleviate pressure on stormwater drainage systems, and decrease the risk of localized flooding.

B. Surface Texture and Slope

Pavement selection also considers surface texture and slope, which affect how water flows across the pavement. Textured or open-graded pavements can promote surface infiltration by increasing the surface area available for water absorption. Additionally, selecting pavements with a permeable base or incorporating features like crowned surfaces or longitudinal slopes can encourage water to drain away from vulnerable areas, reducing the risk of ponding and flooding.

C. Water Storage and Detention

Certain pavement types, such as porous asphalt or permeable pavers with underlying stone reservoirs, can provide temporary storage and detention of stormwater. These pavements retain water during rainfall events and gradually release it into the ground or drainage systems, helping to attenuate peak flows and reduce the likelihood of downstream flooding.

D. Integration with Green Infrastructure

Pavement selection can be integrated with green infrastructure practices to enhance flood mitigation efforts. Combining permeable pavements with features like vegetated swales, bioretention areas, or rain gardens helps create a comprehensive stormwater management system that captures, treats, and infiltrates runoff before it enters waterways, reducing the risk of flooding and improving water quality.

E. Maintenance Considerations

Different pavement materials require varying levels of maintenance to ensure their continued effectiveness in managing stormwater. Proper maintenance, including regular cleaning, debris removal, and sedimentation management, is essential for preserving permeability and preventing clogging, which can compromise drainage capacity and increase flood risk.

F. Long-Term Resilience

Selecting durable, resilient pavement materials that withstand weathering and heavy traffic ensures the long-term effectiveness of flood mitigation measures. Pavement selection should consider factors such as lifespan, resistance to erosion, and ability to withstand freeze-thaw cycles to minimize the need for frequent repairs or replacement, which can disrupt stormwater management efforts and compromise flood resilience.

By carefully considering factors such as permeability, surface texture, slope, water storage capacity, integration with green infrastructure, maintenance requirements, and long-term resilience, pavement selection can play a critical role in mitigating flooding and building more sustainable, resilient communities.

4. Design Considerations

- Work with engineers and designers to develop a permeable pavement system that maximizes stormwater infiltration and minimizes runoff.

- Incorporate proper surface slopes and drainage channels to direct water towards infiltration points and prevent ponding.

- Consider factors such as surface slope, drainage channels, and infiltration rates to optimize performance.

A. Retention and Detention Basins

Designing retention and detention basins allows for the temporary storage of stormwater, which helps attenuate peak flows and reduces the risk of downstream flooding. These basins can be integrated into green spaces, parks, or other landscaped areas to provide aesthetic and recreational benefits while also serving as flood control measures.

B. Swales and Bioswales

Incorporating swales and bioswales into the landscape design helps capture and treat stormwater runoff. These vegetated channels or depressions allow water to infiltrate into the ground while removing pollutants and sediments, thereby reducing the volume and velocity of runoff and mitigating flooding.

C. Permeable Pavements

Integrating permeable pavements into the design helps reduce surface runoff by allowing rainwater to infiltrate into the ground. Permeable pavements can be strategically placed in parking lots, sidewalks, driveways, and other impermeable surfaces to minimize the amount of runoff reaching storm drains and waterways, thereby mitigating flooding.

D. Green Roofs

Designing buildings with green roofs can help mitigate flooding by absorbing and retaining rainwater. Green roofs act as natural sponges, reducing the volume of runoff and delaying its release into the drainage system. This helps alleviate pressure on stormwater infrastructure and reduces the risk of urban flooding.

E. Rain Gardens

Incorporating rain gardens into the landscape design provides an attractive and effective way to manage stormwater. These shallow, vegetated depressions capture and absorb rainwater, allowing it to infiltrate into the soil and recharge groundwater. Rain gardens also enhance biodiversity, improve water quality, and reduce the risk of localized flooding.

F. Retention of Natural Features

Preserving and enhancing natural features such as wetlands, floodplains, and riparian buffers helps mitigate flooding by providing natural storage and conveyance areas for stormwater. Designing developments to avoid encroachment on these natural features helps maintain their flood control functions and supports ecological resilience.

G. Sustainable Drainage Systems (SuDS)

Designing with SuDS principles involves integrating a range of drainage techniques that mimic natural processes to manage stormwater. SuDS features such as permeable surfaces, infiltration trenches, and constructed wetlands help reduce runoff volume, improve water quality, and mitigate flooding in urban areas.

H. Community Engagement and Education

Designing flood mitigation measures with community engagement and education in mind helps build awareness and support for sustainable stormwater management practices. Engaging stakeholders in the design process fosters a sense of ownership and promotes behaviors that reduce flood risk, such as proper maintenance of drainage infrastructure and responsible stormwater management on private properties.

By incorporating these design considerations into urban planning and development projects, communities can effectively mitigate flooding, enhance resilience to extreme weather events, and create more sustainable and livable environments.

5. Subbase Preparation

- Prepare the subbase beneath the permeable pavement to enhance drainage and facilitate water infiltration.

- Use aggregate materials with high porosity and permeability to provide storage space for stormwater and promote infiltration into the underlying soil.

Subbase preparation plays a significant role in mitigating flooding by enhancing the effectiveness of permeable pavements and other stormwater management systems. Here's how subbase preparation contributes to flood mitigation

    A. Improving Infiltration

Proper subbase preparation ensures that the ground beneath permeable pavements is adequately prepared to facilitate infiltration of stormwater. This involves assessing soil conditions, addressing any compaction or contamination issues, and providing sufficient drainage capacity to allow water to percolate into the ground. By enhancing infiltration rates, subbase preparation helps reduce surface runoff and alleviate flooding by directing water into the soil instead of allowing it to accumulate on impermeable surfaces.

B. Maintaining Permeability

Subbase preparation helps maintain the long-term permeability of permeable pavements by preventing compaction and sedimentation. Compaction of the subbase layer can reduce pore space and impede water infiltration, diminishing the effectiveness of the pavement in managing stormwater. Proper subbase preparation includes using suitable aggregate materials, ensuring adequate thickness, and compacting the subbase layer to the appropriate density to support the load-bearing requirements of the pavement while preserving permeability.

C. Preventing Clogging

Subbase preparation aims to minimize the risk of clogging within the pavement structure, which can impede drainage and reduce flood mitigation effectiveness. By providing adequate separation between the subbase and underlying soils, installing geotextile or geogrid reinforcements, and incorporating appropriate filter layers, subbase preparation helps prevent the intrusion of fines and sediment into the pavement structure, maintaining its permeability and drainage capacity over time.

D. Enhancing Structural Stability

Properly prepared subbase layers provide a stable foundation for permeable pavements, reducing the risk of pavement failure and erosion during heavy rainfall events. Subbase preparation involves ensuring uniformity of material properties, adequate compaction, and proper grading to support the structural integrity of the pavement system under various loading conditions. By enhancing structural stability, subbase preparation helps maintain the functionality of permeable pavements in managing stormwater and mitigating flooding.

E. Facilitating Maintenance

Subbase preparation facilitates routine maintenance of permeable pavements, which is essential for ensuring their continued effectiveness in flood mitigation. A well-prepared subbase layer allows for easier access to the pavement structure for inspection, cleaning, and maintenance activities. Regular maintenance, such as vacuuming, pressure washing, and sediment removal, helps prevent clogging and sedimentation, prolonging the service life of permeable pavements and maximizing their flood mitigation benefits.

F. Subbase preparation is a critical component of stormwater management and flood mitigation efforts, as it ensures the proper functioning and longevity of permeable pavements and other infiltration-based systems. By enhancing infiltration, maintaining permeability, preventing clogging, enhancing structural stability, and facilitating maintenance, subbase preparation contributes to more effective and sustainable flood mitigation strategies.

6. Installation Practices

- Ensure that permeable pavement systems are installed according to manufacturer specifications and industry best practices.

- Pay close attention to surface grading, joint spacing, and compaction to achieve optimal performance and longevity.

- Prepare the site by excavating the existing pavement and ensuring proper subgrade preparation.

- Pay attention to surface grading and joint spacing to facilitate water infiltration and prevent ponding.

Installation practices play a crucial role in mitigating flooding by ensuring the effectiveness and longevity of stormwater management infrastructure. Here's how proper installation practices can contribute to flood mitigation

A. Proper Grading and Sloping

During installation, proper grading and sloping of surfaces direct stormwater runoff away from vulnerable areas and toward appropriate drainage systems. This prevents water from pooling or accumulating in low-lying areas, reducing the risk of localized flooding.

B. Ensuring Adequate Drainage

Proper installation includes the installation of adequate drainage infrastructure such as catch basins, culverts, and underground pipes to convey stormwater away from developed areas. By providing a well-designed drainage network, installation practices help prevent the accumulation of water on pavements and surrounding surfaces, mitigating the risk of flooding.

C. Maintaining Surface Permeability

For permeable pavements and other infiltration-based systems, proper installation practices ensure the preservation of surface permeability. This involves using suitable aggregate materials, compacting the subbase layer to the appropriate density, and avoiding the use of heavy machinery that could damage the pavement structure. By maintaining permeability, installation practices facilitate the infiltration of stormwater into the ground, reducing surface runoff and alleviating flooding.

D. Effective Erosion Control Measures

Installation practices should include effective erosion control measures to prevent sedimentation and siltation of waterways. This may involve installing erosion control blankets, silt fences, or vegetative stabilization measures to stabilize soil and prevent sediment runoff into adjacent water bodies. By minimizing erosion and sedimentation, installation practices help preserve water quality and maintain the capacity of drainage systems to convey stormwater effectively, reducing the risk of flooding.

E. Attention to Detail and Quality Assurance

Proper installation practices require attention to detail and adherence to industry standards and best practices. This includes proper compaction of soil and base materials, precise placement of drainage infrastructure, and thorough quality assurance measures to ensure the integrity and functionality of stormwater management systems. By prioritizing quality installation, practitioners can minimize the risk of system failures, optimize performance, and enhance resilience to flooding events.

F. Training and Certification Program

Training and certification programs for installation professionals ensure that practitioners have the knowledge and skills necessary to implement stormwater management infrastructure effectively. By providing training on proper installation techniques, equipment operation, and maintenance practices, these programs help improve the quality and consistency of installations, ultimately enhancing flood mitigation efforts.

G. Community Engagement and Education

Effective installation practices should include community engagement and education to raise awareness about the importance of stormwater management and flood mitigation. By involving stakeholders in the installation process and providing guidance on proper maintenance and stewardship of stormwater infrastructure, installation practices can empower communities to actively participate in flood resilience efforts.

By incorporating these installation practices into stormwater management projects, practitioners can enhance the effectiveness and resilience of infrastructure to mitigate flooding and protect communities from the adverse impacts of stormwater runoff.

7. Integration with Drainage Infrastructure

- Integrate permeable pavements with existing drainage systems, such as stormwater pipes, catch basins, and infiltration basins, to manage excess runoff effectively.

- Coordinate with local authorities and utility providers to ensure proper connection and discharge of infiltrated stormwater.

- Direct collected water towards infiltration points or storage areas to reduce the burden on drainage systems.

Integration with drainage infrastructure is crucial for effective flood mitigation as it helps manage stormwater runoff efficiently. Here's how integration with drainage infrastructure can mitigate flooding:

A. Directing Runoff

Integrating stormwater management systems with existing drainage infrastructure, such as storm drains, culverts, and channels, helps direct runoff away from vulnerable areas. By channeling stormwater into designated conveyance systems, integration ensures that excess water is safely transported to appropriate discharge points, reducing the risk of flooding in urban and developed areas.

B. Capacity Enhancement

Integrating stormwater management practices with drainage infrastructure allows for the enhancement of drainage capacity to accommodate increased runoff volumes during heavy rainfall events. By incorporating features such as larger pipes, expanded detention basins, or additional storage reservoirs, integration helps prevent system overload and reduces the likelihood of drainage system failures that can lead to flooding.

C. Green Infrastructure Integration

Integration with drainage infrastructure enables the incorporation of green infrastructure practices, such as rain gardens, bioswales, and vegetated retention ponds, into the urban landscape. These nature-based solutions help capture, treat, and infiltrate stormwater runoff, reducing the burden on traditional drainage systems and mitigating the risk of flooding by slowing down the flow of water and promoting infiltration into the ground.

D. Infiltration and Detention

Integration facilitates the implementation of infiltration and detention practices within drainage infrastructure to mitigate flooding. By incorporating permeable pavements, infiltration trenches, and underground storage chambers into the design of stormwater conveyance systems, integration allows for the temporary storage and gradual release of stormwater, reducing peak flows and minimizing the risk of localized flooding downstream.

E. Multi-Functional Design

Integration with drainage infrastructure promotes the development of multi-functional stormwater management systems that serve additional purposes beyond flood mitigation. By combining flood control measures with features such as recreational amenities, wildlife habitat enhancement, and aesthetic enhancements, integration maximizes the social, environmental, and economic benefits of stormwater infrastructure while effectively reducing flood risk.

F. Climate Resilience Planning

Integration with drainage infrastructure supports climate resilience planning by incorporating adaptive measures that address the anticipated impacts of climate change, such as increased precipitation, sea level rise, and more frequent and intense storms. By designing resilient drainage systems that can withstand extreme weather events and accommodate changing hydrological conditions, integration helps communities adapt to future flood risks and build long-term resilience.

G. Community Engagement and Stakeholder Collaboration

Integration with drainage infrastructure involves engaging stakeholders and collaborating with local communities to identify flood risk areas, prioritize mitigation measures, and implement effective stormwater management strategies. By involving residents, businesses, and other stakeholders in the planning and decision-making process, integration fosters community ownership and support for flood mitigation efforts, ultimately enhancing the success and sustainability of drainage infrastructure projects.

By integrating stormwater management practices with drainage infrastructure, communities can effectively mitigate flooding, reduce the impact of stormwater runoff on the built environment, and enhance overall resilience to extreme weather events.

8. Maintenance Regimen

- Develop a proactive maintenance plan to preserve the functionality and longevity of the permeable pavement system.

- Regularly inspect the pavement surface for debris, sediment accumulation, and signs of clogging.

- Implement routine maintenance activities, such as vacuuming, power washing, and occasional surface rejuvenation, to restore permeability and prevent performance degradation.

- Establish a regular maintenance schedule to ensure the longevity and effectiveness of the permeable pavement system.

- Clean the surface as needed using vacuuming, power washing, or other suitable methods to maintain permeability.

A well-implemented maintenance regimen is critical in mitigating flooding by ensuring the functionality and efficiency of stormwater management infrastructure. Here's how maintenance can help mitigate flooding

  A. Preventing Clogging and Blockages

Regular maintenance, such as cleaning catch basins, storm drains, and culverts, helps prevent debris buildup and blockages that can impede the flow of stormwater. By ensuring that drainage infrastructure remains clear and free-flowing, maintenance reduces the risk of localized flooding caused by blocked drainage systems.

B. Preserving Permeability

For permeable pavements and infiltration-based stormwater management systems, maintenance activities such as vacuuming, power washing, and sediment removal help preserve surface permeability. By preventing the accumulation of sediment, debris, and pollutants within pavement pores, maintenance ensures that water can infiltrate into the ground effectively, reducing surface runoff and mitigating flooding.

C. Inspecting and Repairing Infrastructure

Routine inspections of stormwater management infrastructure allow for the early detection of structural damage, erosion, or deterioration that could compromise system performance. By promptly repairing leaks, cracks, or other defects in drainage pipes, culverts, and retention basins, maintenance helps maintain the integrity and functionality of infrastructure, reducing the risk of system failures and associated flooding.

D. Clearing Vegetation and Overgrowth

Vegetation management is essential for preventing the obstruction of drainage channels, swales, and culverts by overgrown vegetation. Regular trimming and clearing of vegetation along waterways and drainage paths help maintain flow capacity and prevent blockages that can lead to localized flooding during heavy rainfall events.

E. Monitoring and Managing Sedimentation

Sedimentation management is crucial for preserving the capacity and efficiency of stormwater conveyance systems. Regular sediment removal from retention basins, detention ponds, and sediment traps helps prevent the accumulation of sediment, silt, and debris that can reduce storage volume and impede flow, leading to increased flood risk downstream.

F. Adapting to Changing Conditions

Maintenance regimens should be adaptable to changing weather patterns, land use changes, and other factors that may influence flood risk. By incorporating flexible scheduling, responsive monitoring, and adaptive management practices, maintenance programs can effectively address evolving challenges and mitigate flood risk in a dynamic environment.

G. Community Engagement and Participation

Engaging community members in maintenance activities promotes stewardship and ownership of stormwater management infrastructure. By organizing volunteer clean-up events, providing educational materials on stormwater best practices, and encouraging residents to report drainage issues, maintenance regimens can leverage community support to enhance flood mitigation efforts and promote a culture of resilience.

By implementing a comprehensive maintenance regimen that includes regular inspection, cleaning, repair, vegetation management, sedimentation control, and community engagement, communities can effectively mitigate flooding and ensure the long-term functionality and resilience of stormwater management infrastructure.

9. Community Engagement

- Educate residents, property owners, and stakeholders about the benefits of permeable pavements for stormwater management and flood mitigation.

- Foster community support for permeable pavement projects through outreach efforts, workshops, and demonstrations.

- Encourage community involvement in the planning and implementation process to foster support and awareness.

Community engagement plays a crucial role in mitigating flooding by fostering collaboration, raising awareness, and empowering residents to take proactive measures to manage stormwater effectively.

   A. Raising Awareness

Community engagement initiatives raise awareness about the causes and impacts of flooding, helping residents understand how their actions can contribute to or alleviate flood risk. By providing information about stormwater management practices, floodplain mapping, and emergency preparedness, community engagement efforts empower residents to make informed decisions and take actions to reduce their vulnerability to flooding.

   B. Promoting Best Practices

Community engagement programs promote best practices for stormwater management, such as reducing impervious surfaces, installing rain gardens, and properly maintaining drainage infrastructure. By educating residents about the benefits of green infrastructure, permeable pavements, and other sustainable drainage practices, community engagement initiatives encourage widespread adoption of flood mitigation measures at the individual and neighborhood level.

C. Encouraging Participation

Engaging residents in the planning, design, and implementation of flood mitigation projects encourages participation and fosters a sense of ownership and responsibility for community resilience. By involving residents in decision-making processes, soliciting feedback, and incorporating local knowledge and expertise, community engagement initiatives ensure that flood mitigation efforts are responsive to community needs and priorities.

D. Building Partnerships

Community engagement initiatives facilitate collaboration and partnerships between residents, local government agencies, non-profit organizations, and other stakeholders involved in flood mitigation efforts. By bringing together diverse perspectives, resources, and expertise, community engagement fosters collective action and enhances the effectiveness of flood resilience initiatives.

E. Empowering Residents

Community engagement empowers residents to take proactive measures to reduce their flood risk and protect their property and communities. By providing access to information, resources, and support, community engagement initiatives enable residents to implement flood mitigation measures on their properties, such as installing rain barrels, redirecting downspouts, or elevating vulnerable structures.

F. Facilitating Communication

Community engagement initiatives provide platforms for communication and dialogue between residents and decision-makers, facilitating the exchange of information, ideas, and feedback. By fostering open communication and transparency, community engagement efforts build trust, enhance accountability, and ensure that flood mitigation strategies reflect community priorities and concerns.

G. Promoting Resilience

Community engagement fosters a culture of resilience by encouraging residents to prepare for and respond to flooding events in proactive and adaptive ways. By promoting emergency preparedness, flood insurance enrollment, and community-based flood monitoring and response efforts, community engagement initiatives help communities build resilience to the impacts of flooding and other natural hazards.

H. Community engagement is essential for building community resilience to flooding by raising awareness, promoting best practices, encouraging participation, building partnerships, empowering residents, facilitating communication, and promoting a culture of resilience. By actively engaging residents in flood mitigation efforts, communities can work together to reduce flood risk, enhance preparedness, and build a more resilient future.

10. Monitoring and Adaptation

- Monitor the performance of the permeable pavement system over time to assess its effectiveness in managing stormwater and reducing flooding.

- Make necessary adjustments or improvements based on observed outcomes and changing conditions.

Monitoring and adaptation are critical components of flood mitigation strategies as they allow for the assessment of effectiveness and adjustment of measures in response to changing conditions. Here's how monitoring and adaptation can mitigate flooding

A. Assessing effectiveness

Monitoring helps evaluate the performance of flood mitigation measures by tracking key indicators such as water levels, rainfall intensity, and floodplain dynamics. By collecting data on the effectiveness of infrastructure, policies, and practices, monitoring enables decision-makers to identify areas of success and areas needing improvement in flood mitigation efforts.

B. Early Warning Systems

Monitoring enables the development and implementation of early warning systems that provide timely alerts about impending floods. By monitoring weather patterns, river levels, and other relevant parameters, early warning systems can give communities and authorities sufficient time to implement flood preparedness measures, evacuate vulnerable areas, and deploy emergency response resources, reducing the impact of flooding on lives and property.

C. Adaptive Management

Adaptation involves adjusting flood mitigation strategies in response to changing conditions, such as increased rainfall, changing land use patterns, or evolving community needs. By adopting an adaptive management approach, decision-makers can continuously review and update flood risk assessments, infrastructure designs, land use policies, and emergency response plans to reflect current and anticipated flood risks and vulnerabilities.

D. Infrastructure Maintenance

Monitoring helps identify maintenance needs and prioritize repairs or upgrades to flood control infrastructure. By regularly inspecting and maintaining drainage systems, levees, dams, and other flood protection structures, authorities can ensure their continued functionality and resilience, reducing the likelihood of system failures and associated flood damage.

E. Environmental Monitoring

Monitoring environmental indicators such as riverine habitats, wetland health, and water quality helps assess the ecological impacts of flooding and flood mitigation measures. By monitoring ecosystem health and biodiversity, decision-makers can incorporate ecological considerations into flood management planning and design, enhancing the resilience of natural systems to flooding and supporting sustainable flood mitigation strategies.

F. Community Feedback and Engagement

Monitoring includes gathering feedback from communities affected by flooding to understand their experiences, concerns, and needs. By engaging with residents, businesses, and other stakeholders, decision-makers can incorporate local knowledge and perspectives into flood mitigation efforts, build trust, and foster collaboration in addressing flood risk and resilience.

G. Research and Innovation

Monitoring provides valuable data for research and innovation in flood risk assessment, prediction, and management. By supporting research initiatives and technological advancements, monitoring helps identify new approaches, tools, and technologies for mitigating flooding, improving flood forecasting accuracy, and enhancing community resilience to flood hazards.

Monitoring and adaptation are essential for effective flood mitigation by providing timely information, supporting informed decision-making, and enabling proactive responses to changing flood risks and conditions. By incorporating monitoring and adaptation into flood management strategies, communities can enhance their resilience to flooding and minimize the impacts of future flood events on lives, livelihoods, and ecosystems. By implementing these steps and integrating permeable pavements into flooding-prone areas with heavy rainfall, communities can enhance stormwater management, reduce runoff, alleviate flooding, and promote sustainable urban development. Collaboration among stakeholders, including government agencies, engineers, designers, and residents, is essential for successful implementation and long-term success.

Sustainable Urban Planning

Plan:

- Adopt water-sensitive urban design principles to minimize impervious surfaces, preserve natural drainage channels, and promote sustainable land use practices that reduce flood risk.

-Sustainable urban planning plays a crucial role in managing flooding in cities.

Action

1. Green Infrastructure

Incorporating green spaces such as parks, green roofs, rain gardens, and permeable pavements can help absorb and store rainwater, reducing the amount of runoff that contributes to flooding. These features also enhance biodiversity and improve air quality. Green infrastructure in sustainable urban planning can mitigate flooding in several ways:

A. Stormwater Management

Green infrastructure features such as permeable pavements, green roofs, and rain gardens help absorb and retain rainwater instead of allowing it to run off into storm drains. By reducing the volume of stormwater runoff, these features help alleviate pressure on drainage systems during heavy rainfall events, thus lowering the risk of flooding.

B. Floodwater Storage

Wetlands, vegetated swales, and constructed ponds are examples of green infrastructure elements that can act as natural storage areas for floodwater. These features temporarily retain excess water during storms, gradually releasing it back into the environment or allowing it to infiltrate into the ground, thereby reducing the peak flow of water in watercourses and minimizing flood risk downstream.

C. Reduced Erosion

Trees, shrubs, and other vegetation within green infrastructure systems help stabilize soil and prevent erosion, particularly along riverbanks and watercourses. By maintaining the integrity of riverbanks and reducing sedimentation, green infrastructure helps maintain natural drainage patterns and prevents the buildup of debris that can exacerbate flooding.

D. Improved Water Quality

Green infrastructure promotes the filtration and purification of stormwater runoff by allowing it to percolate through soil and vegetation before reaching water bodies. This helps remove pollutants, sediments, and contaminants from runoff, improving the overall quality of water in rivers, lakes, and streams. Cleaner watercourses are better able to accommodate increased flow during heavy rainfall events without causing flooding or exacerbating downstream pollution.

E. Urban Heat Island Reduction

Green infrastructure features such as street trees and green spaces help reduce the urban heat island effect by providing shade and evaporative cooling. By moderating temperatures in urban areas, these features can indirectly mitigate flooding by reducing the intensity of convective storms and delaying the onset of runoff, allowing more time for infiltration and absorption of rainwater into the ground.

F. Biodiversity Enhancement

Incorporating green infrastructure into urban areas creates habitats for native flora and fauna, supporting biodiversity and ecosystem resilience. Healthy ecosystems with diverse vegetation are better able to absorb and manage water, regulate hydrological processes, and withstand extreme weather events, thus contributing to overall flood mitigation efforts.

G. Community Benefits

Green infrastructure projects often provide additional benefits to communities beyond flood mitigation, including recreational opportunities, improved air quality, enhanced aesthetics, and increased property values. These co-benefits can help garner public support for sustainable urban planning initiatives and foster a sense of ownership and stewardship among residents, further strengthening community resilience to flooding.

Integrating green infrastructure into sustainable urban planning practices, cities can effectively manage flood risk while promoting environmental sustainability, social equity, and economic vitality.

2. Natural Flood Management

Implementing natural flood management techniques such as restoring wetlands, creating buffer zones along rivers, and reconnecting floodplains can help slow down the flow of water, reduce peak flood levels, and provide additional storage capacity during heavy rainfall events. Natural flood management (NFM) strategies can play a significant role in mitigating flooding within sustainable urban planning frameworks. Here's how:

A. Retention and Infiltration

NFM techniques such as restoring wetlands, reforesting riparian zones, and creating vegetated buffer strips along watercourses help retain and infiltrate rainwater. These natural features act as sponges, absorbing excess water during heavy rainfall events and reducing the volume of runoff that contributes to urban flooding.

B. Reduced Peak Flows

By slowing down the movement of water through the landscape, NFM measures help reduce peak flows in rivers and streams. This can help alleviate pressure on downstream areas and mitigate the risk of flash flooding during intense storms. Techniques such as tree planting, soil conservation, and contouring can also enhance water retention and reduce surface runoff.

C. Floodplain Reconnection

Restoring natural floodplains and reconnecting rivers to their historical floodplain areas allow water to spread out horizontally during floods, reducing the depth and velocity of floodwaters. This approach not only helps protect urban areas from flooding but also enhances habitat connectivity, promotes biodiversity, and improves overall ecosystem health.

D. Channel Naturalization

Naturalizing or restoring rivers and streams to their natural, meandering channels can improve their capacity to accommodate high flows and reduce the risk of bank erosion and channelization. Soft engineering techniques such as adding woody debris, creating riffles and pools, and enhancing streambank vegetation help stabilize riverbanks, increase roughness, and dissipate energy, thereby reducing the erosive power of floodwaters.

E. Sponge Cities Concept

The concept of "sponge cities" involves redesigning urban areas to mimic natural hydrological processes and enhance water absorption and retention capacity. This can include features such as green roofs, permeable pavements, rain gardens, and urban wetlands, which help capture and store rainwater within the urban landscape, reducing runoff and mitigating flooding.

F. Community Engagement

Engaging local communities in NFM initiatives can enhance flood resilience and foster a sense of ownership and stewardship. Community-based projects such as tree planting, river cleanups, and habitat restoration not only contribute to flood risk reduction but also provide opportunities for education, recreation, and social cohesion.

G. Integrated Water Management

Integrating NFM measures into broader water management strategies ensures that flood risk reduction efforts are coordinated and synergistic with other urban planning goals. This may involve collaboration across different sectors, including water resource management, land use planning, environmental conservation, and climate adaptation.

Incorporating natural flood management principles into sustainable urban planning frameworks, cities can enhance their resilience to flooding, promote ecosystem health and biodiversity, and create more livable and sustainable communities for current and future generations.

3. Integrated Water Management

Adopting an integrated approach to water management involves considering the entire water cycle, from rainfall to runoff to discharge. This can involve implementing measures such as sustainable drainage systems (SuDS), rainwater harvesting, and water-sensitive urban design to reduce the volume and impact of stormwater runoff.

Integrated water management (IWM) is a holistic approach that considers the entire water cycle, from precipitation to consumption to disposal, in urban planning and development. Implementing IWM strategies can effectively mitigate flooding in urban areas. Here's how:

A. Stormwater Management

IWM emphasizes the use of sustainable drainage systems (SuDS) to manage stormwater at its source. SuDS include features such as permeable pavements, green roofs, rain gardens, and swales that capture, infiltrate, or store rainwater on-site, reducing the volume and velocity of runoff and alleviating pressure on conventional drainage systems during heavy rainfall events.

B. Flood Risk Assessment and Mapping

Integrating flood risk assessment and mapping into urban planning processes helps identify areas prone to flooding and inform land-use decisions, development regulations, and infrastructure investments. By considering flood risk upfront, planners can avoid development in high-risk areas, preserve natural floodplains, and ensure that new developments are resilient to flooding.

C. Floodplain Management

IWM incorporates floodplain management measures such as zoning regulations, land-use planning, and floodplain preservation to minimize exposure to flood risk and protect lives and property. By avoiding construction in flood-prone areas and preserving natural floodplains, cities can reduce the impact of flooding and minimize damage to infrastructure and communities.

D. Green Infrastructure

Integrating green infrastructure into urban landscapes helps mitigate flooding by promoting natural infiltration, reducing runoff, and enhancing water storage capacity. Green infrastructure features such as parks, wetlands, vegetated swales, and riparian buffers absorb rainwater, slow down its movement, and provide additional floodwater storage, thereby reducing flood risk downstream.

E. Climate Resilience

IWM considers the potential impacts of climate change, such as increased rainfall intensity and sea-level rise, on urban flooding. By incorporating climate resilience measures into planning and design, cities can adapt to changing flood risk dynamics, upgrade infrastructure to withstand extreme weather events, and ensure the long-term sustainability of water management systems.

F. Multi-Stakeholder Collaboration

IWM involves collaboration among various stakeholders, including government agencies, water utilities, urban planners, engineers, developers, and community groups. By fostering partnerships and coordinating efforts across sectors, cities can leverage expertise, resources, and funding to implement comprehensive flood mitigation strategies and maximize effectiveness.

G. Public Education and Awareness

Engaging the public through education and outreach programs raises awareness about flood risk, promotes community preparedness, and encourages behavior change. By empowering residents to take proactive measures to protect themselves and their properties, cities can strengthen resilience to flooding and build a culture of safety and preparedness.

By integrating these principles into sustainable urban planning processes, cities can effectively mitigate flooding, reduce vulnerability to climate change impacts, and create safer, more resilient communities for all residents.

4. Floodplain Management

Avoiding construction in flood-prone areas and preserving natural floodplains can help reduce the risk of flooding and minimize damage to properties and infrastructure. Zoning regulations and land-use planning should take into account flood risk and ensure that development is located in safer areas.

Floodplain management is a key component of sustainable urban planning that focuses on reducing the risk of flooding and minimizing its impact on communities. Here's how floodplain management can mitigate flooding within sustainable urban planning frameworks:

A. Zoning Regulations

Implementing zoning regulations that restrict or prohibit development in flood-prone areas helps prevent the construction of homes, businesses, and infrastructure in high-risk zones. By avoiding development in floodplains, cities can reduce exposure to flood hazards and minimize potential damage to property and infrastructure.

B. Floodplain Mapping

Conducting detailed floodplain mapping allows urban planners to identify areas at risk of flooding and assess the potential impact of inundation. This information helps inform land-use decisions, infrastructure planning, and emergency preparedness efforts, enabling cities to prioritize floodplain management strategies and allocate resources effectively.

C. Floodplain Preservation

Preserving natural floodplains and riparian areas helps maintain the natural storage capacity of river systems and reduces the risk of downstream flooding. Protecting wetlands, forests, and other natural features within floodplains provides additional space for floodwaters to spread out horizontally, attenuating peak flows and mitigating flood risk for adjacent communities.

D. Floodplain Setbacks

Establishing setback requirements that mandate minimum distances between development and water bodies helps minimize exposure to flood risk and preserve the integrity of floodplains. By maintaining buffer zones along rivers, streams, and coastal areas, cities can reduce the likelihood of property damage, erosion, and loss of habitat during flood events.

E. Floodplain Restoration

Restoring degraded floodplains and reestablishing natural hydrological processes can enhance flood resilience and ecosystem health. Projects such as floodplain reforestation, wetland restoration, and stream channel realignment help improve water retention, increase infiltration, and reduce the erosive power of floodwaters, thereby reducing flood risk and enhancing ecological resilience.

F. Floodplain Regulations and Building Codes

Enforcing regulations and building codes that require flood-resistant construction standards helps mitigate flood damage and protect public safety. Measures such as elevating structures above base flood elevations, using flood-resistant materials, and installing floodproofing measures help reduce vulnerability to flooding and minimize the financial and social costs associated with flood damage.

G. Floodplain Management Plans

Developing comprehensive floodplain management plans that integrate land-use planning, infrastructure development, emergency preparedness, and community engagement facilitates coordinated action and ensures the implementation of effective flood mitigation strategies. These plans identify goals, objectives, and actions for reducing flood risk, enhancing resilience, and protecting lives and property in flood-prone areas.

Incorporating these floodplain management strategies into sustainable urban planning frameworks, cities can effectively mitigate flooding, reduce vulnerability to climate change impacts, and create safer, more resilient communities for current and future generations.

5. Climate Resilience

Designing infrastructure and buildings to withstand extreme weather events, including flooding, is essential for building climate resilience. This can involve elevating structures, using flood-resistant materials, and designing infrastructure with redundancy and flexibility to adapt to changing conditions.

6. Community Engagement and Education

Engaging with local communities and raising awareness about flood risk can help empower residents to take proactive measures to protect themselves and their properties. Community-based initiatives such as flood preparedness programs, emergency response training, and neighborhood flood watch schemes can strengthen resilience at the grassroots level.

7. Cross-Sector Collaboration

Effective flood management requires collaboration across different sectors and stakeholders, including government agencies, urban planners, engineers, environmental organizations, and local communities. Integrated planning and decision-making processes can help ensure that flood risk reduction measures are coordinated and implemented effectively.

8. Long-Term Planning and Adaptation

Given the increasing frequency and intensity of extreme weather events due to climate change, long-term planning and adaptation strategies are essential for managing flood risk in urban areas. This may involve updating flood maps, revising building codes and regulations, and investing in infrastructure upgrades to improve resilience to future flood events.

Incorporating these principles into urban planning and development practices, cities can enhance their resilience to flooding while promoting sustainable and livable communities for current and future generations.

Smart Drainage Systems

Plan:

- Implement smart drainage systems equipped with sensors and real-time monitoring technology to optimize drainage operations, detect blockages, and manage stormwater flow more effectively during heavy rainfall. Smart drainage systems utilize advanced technologies such as sensors, telemetry, and real-time monitoring to optimize drainage operations and mitigate flooding.

Action

1. Real-Time Monitoring

    Smart drainage systems continuously monitor water levels, flow rates, and weather conditions in real-time. This data is transmitted to a central control system, allowing operators to assess current conditions and respond quickly to changing circumstances.

    Real-time monitoring in smart drainage systems involves the integration of sensors, data collection, and analysis technologies to continuously monitor and manage water levels, flow rates, and other relevant parameters within drainage networks. Here's how real-time monitoring contributes to the effectiveness of smart drainage systems:

    A. Early Warning Systems
    Real-time monitoring enables the detection of sudden changes in water levels and flow rates, providing early warnings of potential flooding events. This allows authorities to implement preventive measures, such as activating pumps

or deploying flood barriers, to mitigate flood risk and protect infrastructure and communities.

B. Data Collection

Advanced sensors installed throughout the drainage network collect real-time data on various parameters, including rainfall intensity, water levels, flow velocities, and water quality. This data provides valuable insights into the current conditions of the drainage system, helping authorities make informed decisions about flood management strategies and infrastructure maintenance.

C. Dynamic Control

Real-time monitoring facilitates dynamic control of drainage infrastructure, allowing for adjustments to be made in response to changing conditions. For example, automated valves and gates can be controlled based on real-time data to optimize water flow within the drainage network, preventing localized flooding and reducing the risk of overflows.

D. Flood Forecasting

By analyzing real-time data alongside historical records and weather forecasts, smart drainage systems can generate flood forecasts and assess the likelihood and severity of potential flooding events. This enables authorities to anticipate flood risk, allocate resources effectively, and implement preemptive measures to minimize damage and disruption.

E. Remote Monitoring and Management

Real-time monitoring systems can be accessed remotely via centralized control centers or mobile applications, allowing authorities to monitor the status of the drainage network in real-time and respond promptly to emerging issues. Remote management capabilities improve operational efficiency and reduce the need for physical inspections, saving time and resources.

F. Data Analytics

Real-time monitoring systems employ advanced data analytics techniques, such as machine learning and predictive modeling, to analyze large volumes of data and identify patterns, trends, and anomalies within the drainage network. This enables authorities to optimize drainage system performance, identify potential sources of blockages or leaks, and prioritize maintenance activities.

G. Integration with Decision Support Systems

Real-time monitoring data can be integrated with decision support systems and visualization tools to provide actionable insights to decision-makers. Interactive dashboards, maps, and reports facilitate data-driven decision-making and enhance situational awareness during flood events, enabling authorities to coordinate response efforts effectively.

Real-time monitoring plays a crucial role in enhancing the resilience and efficiency of smart drainage systems, enabling proactive flood management, reducing the impact of flooding on urban areas, and ensuring the long-term sustainability of water infrastructure.

2. Predictive Modeling

By analyzing historical data and weather forecasts, smart drainage systems can predict potential flooding events with greater accuracy. This enables proactive measures to be taken, such as adjusting flow rates, activating pumps, or deploying flood barriers, to prevent or minimize flooding.

Predictive modeling in smart drainage systems involves the use of mathematical algorithms and computational techniques to forecast future flood events, assess flood risk, and optimize drainage system operations. Here's how predictive modeling contributes to the effectiveness of smart drainage systems

A. Flood Forecasting
Predictive models analyze historical data on rainfall, water levels, soil moisture, land use, and other relevant factors to forecast future flood events. By simulating various scenarios and predicting how different factors will interact, these models can provide advance warning of potential flooding, allowing authorities to implement preventive measures and mitigate flood risk.

B. Risk Assessment
Predictive modeling assesses the vulnerability of urban areas to flooding by identifying areas at high risk based on factors such as topography, land use, drainage infrastructure, and historical flood data. This information helps authorities prioritize flood management strategies, allocate resources effectively, and develop targeted interventions to reduce flood risk in vulnerable areas.

C. Infrastructure Planning and Design
Predictive modeling informs the planning and design of drainage infrastructure by simulating the performance of different system configurations under varying conditions. By optimizing the layout, capacity, and operation of drainage networks based on predictive models, authorities can minimize flood risk, reduce infrastructure costs, and ensure the resilience of urban water systems.

D. Operational Optimization

Predictive models optimize the operation of drainage infrastructure by predicting future water levels, flow rates, and flood volumes based on real-time data and forecasted weather conditions. By dynamically adjusting control strategies, such as opening or closing flood gates and adjusting pump settings, predictive models help maximize the efficiency and effectiveness of drainage systems in managing flood risk.

E. Emergency Response Planning

Predictive modeling supports emergency response planning by providing decision-makers with actionable insights into potential flood scenarios and their potential impacts on infrastructure, communities, and the environment. By simulating different response strategies and their outcomes, predictive models enable authorities to develop robust emergency plans, coordinate response efforts, and minimize the consequences of flood events.

F. Climate Change Adaptation

Predictive modeling helps cities adapt to the impacts of climate change by assessing future flood risk under different climate scenarios. By projecting how climate change will affect rainfall patterns, sea levels, and extreme weather events, predictive models inform long-term adaptation strategies, infrastructure investments, and land-use planning decisions to build resilience to future flood risk.

G. Continuous Improvement

Predictive modeling relies on continuous monitoring and updating of data to improve accuracy and reliability over time. By incorporating new data sources, refining algorithms, and validating model outputs against observed data, predictive models evolve and improve their predictive capabilities, enabling more effective flood risk management and decision-making.

Predictive modeling plays a critical role in enhancing the resilience, efficiency, and effectiveness of smart drainage systems, enabling proactive flood management, reducing the impact of flooding on urban areas, and ensuring the sustainable management of water resources.

3. Automated Controls
Smart drainage systems are equipped with automated controls that can adjust drainage infrastructure settings based on preset criteria or sensor inputs. For example, pumps can be activated automatically when water levels rise above a certain threshold, helping to prevent localized flooding in low-lying areas.

Automated controls in smart drainage systems involve the use of sensors, actuators, and control algorithms to automate the operation of drainage infrastructure in response to changing conditions. Here's how automated controls enhance the effectiveness of smart drainage systems

A. Real-Time Monitoring
Automated controls rely on real-time data from sensors installed throughout the drainage network to monitor water levels, flow rates, and other relevant parameters. By continuously monitoring these variables, automated systems can detect changes in drainage conditions and respond promptly to prevent flooding or system failures.

B. Dynamic Adjustment

Automated controls enable dynamic adjustment of drainage infrastructure in response to changing conditions. For example, automated valves and gates can be opened or closed based on real-time data to regulate water flow within the drainage network, optimizing drainage performance and minimizing the risk of overflows or blockages.

C. Predictive Analytics

Automated controls use predictive analytics algorithms to forecast future drainage conditions based on historical data, weather forecasts, and other relevant factors. By predicting potential flooding events in advance, automated systems can proactively adjust drainage operations to mitigate flood risk and protect infrastructure and communities.

D. Remote Monitoring and Management

Automated controls allow for remote monitoring and management of drainage infrastructure from centralized control centers or mobile devices. This enables authorities to monitor the status of the drainage network in real-time, receive alerts about potential issues, and take corrective actions remotely, improving operational efficiency and reducing response times.

E. Integration with Decision Support Systems

Automated controls integrate with decision support systems and visualization tools to provide actionable insights to decision-makers. Interactive dashboards, maps, and reports facilitate data-driven decision-making and enhance situational awareness during flood events, enabling authorities to coordinate response efforts effectively.

F. Adaptive Operation

Automated controls enable adaptive operation of drainage infrastructure based on changing conditions and user-defined criteria. For example, automated systems can adjust pumping rates, open or close flood gates, or redirect flow based on predefined rules or algorithms, optimizing system performance and minimizing energy consumption.

G. Fault Detection and Diagnostics

Automated controls include built-in fault detection and diagnostics capabilities to identify and address issues in real-time. By continuously monitoring equipment status and performance metrics, automated systems can detect malfunctions, leaks, or blockages early and alert operators to take corrective actions, reducing downtime and maintenance costs.

Automated controls play a critical role in enhancing the resilience, efficiency, and effectiveness of smart drainage systems, enabling proactive flood management, reducing the impact of flooding on urban areas, and ensuring the sustainable management of water resources.

4. Remote Monitoring and Control

Smart drainage systems allow for remote monitoring and control of drainage infrastructure from a centralized location. This enables operators to respond to flooding events quickly, even in inaccessible or hazardous environments, reducing response times and minimizing flood damage.

Remote monitoring and control in smart drainage systems utilize advanced technologies to oversee and manage drainage infrastructure from a centralized location, often through digital interfaces or mobile applications. Here's how remote monitoring and control enhance the functionality of smart drainage systems

A. Real-Time Data Monitoring

Remote monitoring systems continuously collect real-time data from sensors installed throughout the drainage network, including water levels, flow rates, rainfall intensity, and water quality parameters. This data provides insights into the current conditions of the drainage system, allowing operators to identify issues and respond promptly.

B. Alerts and Notifications

Remote monitoring systems generate alerts and notifications when predefined thresholds or anomalies are detected in the data. For example, operators may receive alerts for high water levels, pump failures, or blockages in the drainage network, enabling them to take immediate action to prevent flooding or system failures.

C. Visualization and Reporting

Remote monitoring systems provide visualization tools and reports that display real-time data in a user-friendly format. Dashboards, maps, and graphs allow operators to visualize the status of the drainage network, track trends over time, and identify areas of concern, facilitating informed decision-making and resource allocation.

D. Remote Control and Automation

Remote control capabilities allow operators to adjust and control drainage infrastructure remotely from a centralized control center or mobile device. For example, operators can remotely open or close flood gates, adjust pump settings, or redirect flow within the drainage network based on real-time data and operational requirements.

E. Optimized Operations

Remote monitoring and control systems optimize the operation of drainage infrastructure by dynamically adjusting settings and operations in response to changing conditions. For example, pumps may be activated or deactivated automatically based on water levels or flow rates, maximizing efficiency and minimizing energy consumption.

F. Predictive Analytics

Remote monitoring systems use predictive analytics algorithms to forecast future drainage conditions and anticipate potential issues before they occur. By analyzing historical data, weather forecasts, and other relevant factors, predictive models can identify trends and patterns, enabling operators to proactively address issues and mitigate risks.

G. Data Security and Privacy

Remote monitoring systems prioritize data security and privacy to protect sensitive information from unauthorized access or cyber threats. Encryption protocols, authentication mechanisms, and access controls ensure that data is transmitted and stored securely, safeguarding the integrity and confidentiality of the information.

Remote monitoring and control systems play a crucial role in enhancing the efficiency, reliability, and resilience of smart drainage systems, enabling proactive management of drainage infrastructure, reducing the risk of flooding, and ensuring the sustainable management of water resources.

5. Dynamic Flow Management
   Smart drainage systems use dynamic flow management techniques to optimize the flow of stormwater through the drainage network. This may involve adjusting flow rates, redirecting water to alternate routes, or prioritizing drainage in critical areas to prevent bottlenecks and reduce the risk of flooding.

6. Integrated Data Analytics
   Smart drainage systems integrate data analytics tools to analyze large volumes of data collected from sensors and other sources. This allows for the identification of trends, patterns, and anomalies that can inform decision-making and improve the efficiency of drainage operations.

7. Resilience and Adaptability
    Smart drainage systems are designed to be resilient and adaptable to changing environmental conditions and future challenges. They can be upgraded and expanded over time to accommodate growth and development, ensuring long-term effectiveness in mitigating flooding.

Smart drainage systems offer a proactive and intelligent approach to flood mitigation, leveraging technology to optimize drainage operations and reduce the risk of flooding in urban areas. By providing real-time monitoring, predictive capabilities, and automated controls, these systems enhance the resilience of communities to extreme weather events and contribute to more effective stormwater management.

Flood control structures

Plan:

- Construct flood control structures such as flood barriers, levees, and detention basins to mitigate the impact of heavy rainfall events and prevent flooding in vulnerable areas.

Flood control structures are engineered infrastructure designed to manage and mitigate the impact of flooding in urban and rural areas. These structures aim to regulate the flow of water, store excess water during flood events, and protect communities, infrastructure, and ecosystems from the adverse effects of flooding.

Action

1. Levees and Floodwalls

Levees and floodwalls are embankments or walls constructed along water bodies such as rivers, lakes, or coastlines to contain floodwaters and prevent them from inundating adjacent areas. Levees are typically earthen structures built parallel to the riverbank, while floodwalls are vertical barriers made of concrete or steel. Levees and floodwalls help protect communities and agricultural land from riverine flooding by confining floodwaters within designated floodplains.

2. Dams and Reservoirs

Dams and reservoirs are large-scale flood control structures that regulate the flow of water in rivers and store excess water during periods of heavy rainfall or snowmelt. By controlling the release of water downstream, dams and reservoirs can attenuate peak flows, reduce the risk of downstream flooding, and provide additional water supply for irrigation, drinking water, and hydropower generation.

3. Detention and Retention Basins

Detention and retention basins are artificial depressions or excavated ponds designed to temporarily store stormwater runoff during heavy rainfall events. Detention basins temporarily detain excess water and release it gradually to downstream channels at a controlled rate, reducing peak flows and minimizing the risk of flash flooding. Retention basins retain water permanently, providing habitat for wildlife and enhancing groundwater recharge.

4. Channel Modifications

Channel modifications involve altering the natural flow characteristics of rivers and streams to improve drainage capacity and reduce the risk of flooding. Techniques such as widening, deepening, or straightening channels, installing erosion control measures, and removing debris or sediment can increase conveyance capacity, reduce flow velocities, and minimize the potential for channel overbanking and flooding.

5. Storm Surge Barriers

Storm surge barriers, also known as tidal barriers or sea gates, are large-scale structures installed in coastal areas to protect against storm surges and tidal flooding. These barriers can be permanently closed during extreme weather events to block incoming tidal waters and prevent coastal inundation, safeguarding coastal communities, infrastructure, and ecosystems from the impacts of storm surge and sea level rise.

6. Urban Drainage Systems

Urban drainage systems include a network of pipes, culverts, and drainage channels designed to convey stormwater runoff from urban areas to receiving water bodies. Flood control structures such as stormwater detention ponds, green infrastructure, and flood gates are integrated into urban drainage systems to manage runoff, reduce the risk of urban flooding, and protect property and infrastructure from inundation.

7. Coastal Protection Structures

Coastal protection structures, such as breakwaters, seawalls, and beach nourishment projects, are built along coastlines to mitigate the effects of erosion, wave action, and coastal flooding. These structures help stabilize shorelines, protect coastal habitats, and reduce the risk of property damage and loss of life during storms and hurricanes.

Flood control structures play a critical role in mitigating flooding and reducing the vulnerability of communities to flood hazards. By regulating water flow, storing excess water, and protecting against inundation, these structures help safeguard lives, property, and ecosystems from the devastating impacts of flooding. However, it's important to consider the potential environmental and social impacts of flood control structures and to implement comprehensive flood management strategies that incorporate nature-based solutions and sustainable land-use planning.

Flood Warning and Management Systems

Plan :

- Develop and deploy flood early warning systems to provide timely alerts to residents and authorities about impending heavy rainfall and potential flood risks, enabling proactive measures to be taken to protect lives and property.

Flood warning and management systems are essential components of flood risk mitigation strategies, providing timely alerts, coordinating emergency response efforts, and facilitating proactive measures to minimize the impact of flooding.

Action

1. Early Warning Systems (EWS)

Flood warning systems utilize real-time data from gauges, sensors, weather forecasts, and hydrological models to monitor water levels, rainfall intensity, and other relevant parameters in rivers, streams, and drainage networks. When predetermined thresholds are exceeded, EWS trigger alerts through various communication channels, such as sirens, text messages, mobile apps, and social media, to warn residents and authorities of impending flood events.

2. Hydrological Modeling

Hydrological models simulate the behavior of water in river basins, watersheds, and drainage systems, allowing authorities to predict river flow, flood extent, and flood risk under different scenarios. By analyzing historical data and forecasted weather conditions, hydrological models provide valuable insights into flood dynamics, enabling authorities to anticipate flood events and implement preventive measures to mitigate risk.

3. Community Engagement and Education

Flood warning and management systems engage with communities through outreach programs, public awareness campaigns, and educational initiatives to raise awareness about flood risk, emergency preparedness, and evacuation procedures. By empowering residents with information and resources, these initiatives promote proactive measures such as floodproofing, evacuation planning, and the purchase of flood insurance, reducing vulnerability to flooding and enhancing community resilience.

4. Integrated Decision Support Systems (DSS)

Integrated DSS combine real-time data, predictive modeling, risk assessment tools, and decision-making frameworks to support informed decision-making and coordination among stakeholders. DSS provide decision-makers with actionable insights into flood risk, enabling them to prioritize response efforts, allocate resources effectively, and implement adaptive strategies to mitigate flooding and minimize damage.

5. Emergency Response Coordination

Flood warning and management systems facilitate coordinated emergency response efforts among government agencies, emergency responders, and other stakeholders. By providing centralized communication platforms, situational awareness tools, and decision support capabilities, these systems enable rapid mobilization of resources, efficient deployment of personnel, and effective coordination of rescue and evacuation operations during flood events.

6. Infrastructure Protection Measures

Flood warning systems integrate with infrastructure protection measures, such as flood barriers, pumps, and levees, to mitigate the impact of flooding on critical infrastructure and assets. By providing advance warning of flood events, these systems enable operators to activate protective measures, deploy temporary flood defenses, and implement emergency protocols to safeguard infrastructure and minimize disruption to essential services.

7. Continuous Improvement and Evaluation

Flood warning and management systems undergo continuous improvement and evaluation to enhance their effectiveness and reliability over time. By analyzing system performance, evaluating response actions, and incorporating lessons learned from past flood events, authorities can identify opportunities for improvement, refine emergency protocols, and invest in upgrades to ensure the resilience and sustainability of flood management efforts.

Flood warning and management systems are essential tools for mitigating flooding and reducing the vulnerability of communities to flood hazards. By providing early warnings, facilitating coordinated response efforts, and promoting proactive measures, these systems help protect lives, property, and ecosystems from the devastating impacts of flooding and contribute to building more resilient and sustainable communities.

Public Education and Awareness

Plan:

- Raise public awareness about flood risks during heavy rainfall events and educate residents on proper stormwater management practices, such as clearing drainage channels and avoiding littering. Migrating from flooding with Community Education and Outreach involves several steps.

Action

1. Assessment
   Understand the current strategies and programs in place for community education and outreach related to flooding. Assess their effectiveness, reach, and impact.
   Public education and awareness assessments play a crucial role in mitigating flooding by informing and empowering individuals and communities to take proactive measures to reduce flood risk. Here's how public education and awareness assessments can help

   A. Understanding of Flood Risks
   Public education and awareness assessments help individuals and communities understand their vulnerability to flooding by providing information about local flood hazards, such as riverine flooding, flash flooding, and coastal flooding. By raising awareness about the causes, impacts, and frequency of floods in their area, education assessments help residents make informed decisions about flood preparedness and mitigation.

   B. Promotion of Preparedness Measures

Public education initiatives promote flood preparedness measures, such as creating emergency evacuation plans, assembling emergency supply kits, and purchasing flood insurance. By providing guidance on how to develop and implement preparedness plans, education assessments empower individuals and communities to take proactive steps to protect themselves and their property before a flood occurs.

C. Encouragement of Mitigation Actions

Public education efforts encourage the adoption of mitigation actions that reduce flood risk and minimize flood damage. This includes promoting practices such as elevating structures above flood levels, retrofitting buildings with flood-resistant materials, and implementing green infrastructure solutions to manage stormwater runoff. By raising awareness about the effectiveness of these mitigation measures, education assessments motivate individuals and communities to invest in flood resilience.

D. Promotion of Responsible Land Use

Public education and awareness assessments advocate for responsible land use practices that reduce flood risk, such as avoiding development in flood-prone areas, preserving natural floodplains, and implementing zoning regulations that restrict construction in high-risk zones. By highlighting the importance of smart land use planning, education assessments help communities avoid future flood losses and minimize the need for costly flood control measures.

E. Encouragement of Community Engagement

Public education initiatives encourage community engagement in flood mitigation efforts by promoting collaboration, communication, and cooperation among residents, businesses, government agencies, and other stakeholders. By fostering a sense of shared responsibility for flood resilience, education assessments create opportunities for collective action and community-led initiatives that address local flood challenges effectively.

F. Dissemination of Emergency Information

Public education assessments ensure that residents have access to timely and accurate information during flood events, including evacuation orders, road closures, and emergency shelter locations. By disseminating emergency information through various channels, such as public announcements, social media, and community outreach events, education assessments help residents stay informed and safe during floods.

G. Promotion of Climate Resilience

Public education efforts raise awareness about the links between flooding and climate change, highlighting the need for adaptation measures to address evolving flood risks. By promoting climate resilience strategies, such as green infrastructure investments, floodplain restoration, and community-based adaptation planning, education assessments help communities prepare for the challenges of a changing climate and build long-term resilience to floods.

H. Public education and awareness assessments are essential components of flood mitigation strategies, providing the knowledge, motivation, and support needed to reduce flood risk, protect lives and property, and build resilient communities. By empowering individuals and communities to understand, prepare for, and respond to flood hazards, education assessments play a vital role in enhancing flood resilience and minimizing the impacts of future flood events.

2. Research

Study successful case studies and best practices in community education and outreach for flooding mitigation and resilience. Identify strategies and approaches that have worked well in similar contexts. Research assessment plays a crucial role in mitigating flooding by providing valuable insights, innovative solutions, and evidence-based strategies to address flood risks effectively. Here's how research assessment contributes to flood mitigation

A. Understanding Flood Dynamics

Research assessment helps deepen our understanding of the complex hydrological processes and factors contributing to flooding, including rainfall patterns, riverine dynamics, coastal processes, and urbanization impacts. By analyzing historical data, conducting field studies, and developing predictive models, research assessment enhances our knowledge of flood dynamics, enabling more accurate flood risk assessments and mitigation planning.

B. Risk Assessment and Mapping

Research assessment supports the development of robust flood risk assessment methodologies and floodplain mapping tools. By integrating data on topography, land use, hydrology, and climate, research assessment helps identify areas at high risk of flooding and assess the potential impacts on communities, infrastructure, and ecosystems. These risk assessments inform land use planning, zoning regulations, and emergency management strategies to minimize flood risk and protect vulnerable populations.

C. Engineering and Infrastructure Design

Research assessment drives innovation in flood control engineering and infrastructure design, leading to the development of more effective and resilient flood mitigation measures. By studying the performance of levees, dams, floodwalls, and drainage systems under various flood scenarios, research assessment informs the design of infrastructure projects that can withstand extreme weather events and reduce flood damage.

D. Green Infrastructure and Natural Flood Management

Research assessment advances our understanding of the role of green infrastructure and natural flood management techniques in mitigating flooding. By studying the effectiveness of practices such as wetland restoration, riparian buffer zones, and permeable pavements, research assessment demonstrates how nature-based solutions can enhance flood resilience, improve water quality, and provide multiple co-benefits for communities and ecosystems.

E. Community-Based Adaptation Strategies

Research assessment supports the development of community-based adaptation strategies that empower local residents and stakeholders to address flood risks collaboratively. By conducting participatory research, stakeholder engagement, and vulnerability assessments, research assessment helps identify community priorities, build social capital, and co-create tailored solutions that reflect local knowledge, values, and needs.

F. Climate Change Adaptation

Research assessment informs climate change adaptation strategies by assessing the potential impacts of climate change on flood risk and developing adaptive responses to mitigate these risks. By modeling future climate scenarios, projecting sea level rise, and assessing changes in precipitation patterns, research assessment helps communities anticipate and prepare for the increased frequency and intensity of floods associated with climate change.

G. Policy Development and Decision-Making

Research assessment provides evidence to inform policy development and decision-making processes related to flood risk management. By synthesizing scientific findings, evaluating policy options, and assessing the costs and benefits of different mitigation strategies, research assessment helps policymakers make informed decisions that prioritize public safety, environmental sustainability, and community resilience.

H. Capacity Building and Knowledge Transfer

Research assessment contributes to capacity building and knowledge transfer initiatives that empower stakeholders to implement effective flood mitigation measures. By disseminating research findings through training programs, workshops, and educational materials, research assessment helps build technical expertise, institutional capacity, and public awareness to support flood resilience efforts at local, national, and global scales.

Research assessment plays a critical role in advancing flood mitigation efforts by generating knowledge, informing policy and practice, fostering innovation, and empowering communities to build resilience to flooding and other climate-related hazards. By investing in research assessment, policymakers, practitioners, and stakeholders can develop evidence-based solutions that reduce flood risk, protect lives and property, and promote sustainable development in flood-prone areas.

3. Engagement

Engage stakeholders including community leaders, local government officials, NGOs, and residents to understand their needs, concerns, and priorities regarding flooding. Community engagement plays a crucial role in mitigating flooding by fostering collaboration, raising awareness, and empowering individuals and communities to take proactive measures to reduce flood risk. Here's how engagement helps mitigate flooding

A. Knowledge Sharing

Engagement facilitates the sharing of knowledge and information about flood risks, mitigation measures, and best practices among residents, businesses, government agencies, and other stakeholders. By exchanging insights, experiences, and lessons learned, engagement builds collective understanding and awareness of flood hazards and resilience strategies within the community.

B. Stakeholder Collaboration

Engagement brings together diverse stakeholders, including residents, businesses, local governments, non-profit organizations, and academia, to collaborate on flood mitigation initiatives. By fostering partnerships and collaboration, engagement leverages collective expertise, resources, and networks to develop holistic, multi-disciplinary solutions that address the complex challenges of flooding effectively.

C. Empowerment and Participation

Engagement empowers individuals and communities to actively participate in flood mitigation efforts by providing opportunities for input, involvement, and decision-making. By engaging residents in planning processes, community meetings, and volunteer activities, engagement fosters a sense of ownership, responsibility, and resilience, empowering individuals to take action to protect themselves and their communities from flooding.

D. Community-Based Solutions

Engagement promotes the development of community-based solutions that reflect local knowledge, values, and priorities. By soliciting input from residents and stakeholders, engagement ensures that flood mitigation measures are tailored to the unique needs, preferences, and circumstances of the community, increasing their relevance, acceptance, and effectiveness.

E. Building Social Capital

Engagement strengthens social capital and community cohesion by fostering trust, cooperation, and mutual support among residents and stakeholders. By creating opportunities for interaction, dialogue, and collaboration, engagement builds social networks and relationships that enhance resilience, facilitate collective action, and promote solidarity in the face of flooding and other hazards.

F. Behavior Change and Resilience

Engagement encourages behavior change and resilience-building actions that reduce flood risk and enhance community preparedness. By providing education, training, and outreach on flood preparedness, mitigation, and recovery, engagement empowers individuals to adopt proactive measures, such as elevating structures, installing flood barriers, and purchasing flood insurance, to protect themselves and their property from flooding.

G. Policy Advocacy and Decision-Making

Engagement informs policy advocacy and decision-making processes related to flood risk management and resilience planning. By amplifying community voices, concerns, and priorities, engagement ensures that flood mitigation policies, plans, and investments align with the needs and aspirations of the community, promoting more inclusive, equitable, and effective approaches to flood resilience.

H. Crisis Response and Recovery

Engagement facilitates crisis response and recovery efforts by mobilizing community resources, support, and solidarity in the aftermath of flooding events. By coordinating volunteer efforts, providing social and emotional support, and advocating for assistance and resources, engagement helps communities recover from flood impacts more quickly and effectively, promoting long-term resilience and recovery.

Engagement is essential for mitigating flooding by building community resilience, promoting collaboration, empowering individuals, and fostering inclusive, participatory approaches to flood risk management. By engaging residents and stakeholders in flood mitigation efforts, communities can enhance their capacity to prepare for, respond to, and recover from flooding events, ultimately reducing the impacts of floods on lives, livelihoods, and ecosystems.

4. Strategy Development

Develop a comprehensive strategy for community education and outreach, including goals, target audiences, messaging, channels, and tactics. Ensure the strategy is aligned with broader flood mitigation and resilience efforts.

Strategy development is crucial for mitigating flooding as it provides a structured approach to identify, prioritize, and implement flood risk reduction measures. Here's how strategy development can help mitigate flooding:

A. Risk Assessment

Strategy development begins with a comprehensive risk assessment to identify flood-prone areas, vulnerable assets, and potential impacts of flooding on communities, infrastructure, and the environment. By analyzing historical data, mapping flood hazards, and assessing exposure and vulnerability, strategy development helps prioritize areas for intervention and target resources where they are most needed to reduce flood risk.

B. Identification of Mitigation Measures

Strategy development involves identifying a range of flood mitigation measures tailored to the specific needs and circumstances of the community. This may include structural measures such as levees, floodwalls, and stormwater detention basins, as well as non-structural measures such as land use planning, zoning regulations, and community education programs. By considering a mix of engineering, policy, and natural solutions, strategy development ensures a comprehensive approach to flood risk reduction that addresses both the causes and consequences of flooding.

C. Cost-Benefit Analysis

Strategy development includes a cost-benefit analysis to evaluate the effectiveness and feasibility of different flood mitigation measures. By assessing the costs of implementation, operation, and maintenance against the expected benefits in terms of reduced flood damages, saved lives, and improved community resilience, strategy development helps prioritize investments and maximize the return on investment in flood risk reduction.

D. Long-Term Planning

Strategy development involves long-term planning to account for future climate change impacts and urbanization trends that may exacerbate flood risk. By incorporating climate projections, demographic changes, and land use scenarios into flood risk assessments and mitigation strategies, strategy development ensures that interventions are robust, adaptive, and sustainable over the long term.

E. Stakeholder Engagement

Strategy development engages a wide range of stakeholders, including residents, businesses, government agencies, and non-profit organizations, in the decision-making process. By soliciting input, feedback, and collaboration from diverse perspectives, strategy development ensures that flood mitigation measures reflect the needs, values, and priorities of the community, increasing their relevance, acceptance, and effectiveness.

F. Policy Development and Implementation
Strategy development leads to the development and implementation of policies, regulations, and incentives that support flood risk reduction efforts. By aligning land use planning, building codes, zoning ordinances, and infrastructure investments with flood mitigation objectives, strategy development creates an enabling environment for flood-resilient development and sustainable growth.

G. Monitoring and Evaluation
Strategy development includes mechanisms for monitoring and evaluating the effectiveness of flood mitigation measures over time. By tracking key performance indicators, assessing progress towards goals, and conducting periodic reviews and updates, strategy development ensures that flood risk reduction efforts remain on track, adaptive, and responsive to changing conditions and priorities.

Strategy development is essential for mitigating flooding by providing a systematic, evidence-based approach to identify, prioritize, and implement flood risk reduction measures. By integrating risk assessment, stakeholder engagement, cost-benefit analysis, long-term planning, policy development, and monitoring and evaluation, strategy development helps communities build resilience to flooding, protect lives and property, and promote sustainable development in flood-prone areas.

5. Partnerships
Forge partnerships with relevant organizations, agencies, and community groups to leverage resources, expertise, and reach. Collaborate on developing and implementing education and outreach initiatives.

Partnerships are crucial in mitigating flooding as they leverage collective expertise, resources, and collaboration to develop and implement effective flood risk reduction measures. Here's how partnerships can help mitigate flooding:

A. Pooling Resources
Partnerships bring together diverse stakeholders, including government agencies, non-profit organizations, community groups, businesses, and academic institutions, to pool resources and expertise for flood mitigation efforts. By combining financial, technical, and human resources, partnerships enable larger-scale and more comprehensive flood risk reduction initiatives than any single entity could achieve alone.

B. Sharing Knowledge and Expertise
Partnerships facilitate the sharing of knowledge, expertise, and best practices among stakeholders with different backgrounds and perspectives. By exchanging insights, lessons learned, and innovative approaches, partnerships enhance collective understanding of flood risks and resilience strategies, leading to more informed decision-making and more effective flood mitigation measures.

C. Collaborative Planning and Decision-Making
Partnerships engage stakeholders in collaborative planning and decision-making processes that prioritize flood risk reduction and resilience-building efforts. By soliciting input,

feedback, and buy-in from diverse perspectives, partnerships ensure that flood mitigation measures reflect the needs, values, and priorities of the community, increasing their relevance, acceptance, and effectiveness.

D. Coordinated Action

Partnerships facilitate coordinated action among stakeholders to implement flood risk reduction measures across multiple jurisdictions and sectors. By aligning policies, regulations, and investments, partnerships ensure that flood mitigation efforts are integrated, cohesive, and mutually reinforcing, maximizing their impact and efficiency in reducing flood risk and enhancing resilience.

E. Building Social Capital

Partnerships strengthen social capital and community resilience by fostering trust, cooperation, and mutual support among stakeholders. By creating opportunities for interaction, dialogue, and collaboration, partnerships build relationships and networks that enhance resilience, facilitate collective action, and promote solidarity in the face of flooding and other hazards.

F. Engaging Vulnerable Communities

Partnerships prioritize the engagement of vulnerable communities in flood risk reduction efforts, ensuring that their voices are heard, and their needs are addressed. By working directly with communities affected by flooding, partnerships build trust, empower residents, and co-create solutions that reflect local knowledge, values, and priorities, leading to more inclusive, equitable, and effective flood mitigation measures.

G. Leveraging Innovation and Technology

Partnerships leverage innovation and technology to develop and implement cutting-edge flood risk reduction measures. By collaborating with research institutions, technology providers, and other stakeholders, partnerships harness new tools, data, and approaches, such as remote sensing, predictive modeling, and real-time monitoring, to enhance flood forecasting, early warning, and response capabilities.

H. Advocacy and Policy Influence

Partnerships advocate for policies, regulations, and investments that support flood risk reduction and resilience-building efforts at local, national, and global levels. By mobilizing stakeholders, raising awareness, and presenting evidence-based arguments, partnerships influence decision-makers and policymakers to prioritize flood resilience in planning, budgeting, and policy-making processes, leading to more effective and sustainable flood mitigation measures.

Partnerships are essential for mitigating flooding by mobilizing collective action, fostering collaboration, and building resilience across communities and sectors. By working together, stakeholders can address the complex challenges of flooding more effectively, reduce risk, and create safer, more resilient communities for all.

6. Develop educational materials

Develop educational materials and resources tailored to different audiences, including information on flood risks, preparedness, response, and recovery. Use a variety of formats such as brochures, fact sheets, videos, and workshops.

   A. Brochures or pamphlets explaining common flood risks, such as flash floods, riverine flooding, and coastal flooding.

B. Infographics illustrating flood preparedness tips, including evacuation routes, emergency contacts, and essential supplies.
C. Online resources, such as interactive maps showing flood-prone areas and educational videos explaining flood safety measures. Flood preparedness guides outlining steps to protect homes and families before, during, and after a flood.
D. Checklists for creating emergency kits, securing valuables, and developing evacuation plans.
E. Workshops or webinars covering topics like flood insurance, floodproofing techniques, and post-flood recovery strategies.
F. Neighborhood meetings or community events focused on flood risk reduction measures and mutual assistance initiatives.
G. Curriculum materials integrating flood risk awareness into science, geography, and social studies lessons.
H. Classroom activities and projects exploring flood-related topics, such as watershed management, climate change impacts, and disaster resilience.
I. Student-led initiatives, such as creating emergency preparedness plans for schools, conducting flood risk assessments in local communities, and organizing flood awareness campaigns.
J. Guest lectures or presentations by experts in hydrology, emergency management, and environmental science.
K. Workplace safety manuals including flood response protocols, evacuation procedures, and business continuity plans.
L. Training sessions for employees on recognizing flood hazards, using emergency equipment, and implementing flood mitigation measures.

M. Business continuity workshops addressing the financial, operational, and legal aspects of flood risk management and recovery.
N. Industry-specific resources and case studies highlighting best practices for protecting assets, minimizing disruption, and resuming operations after a flood.
O. Volunteer training programs on disaster response and recovery, including first aid, search and rescue techniques, and psychological support.
P. Outreach materials encouraging community engagement in flood preparedness activities, such as neighborhood cleanups, floodplain restoration projects, and emergency drills.
Q. Collaboration opportunities with local authorities, emergency services, and nonprofit organizations to develop community-based flood response plans and establish mutual aid networks.
R. Informational sessions and workshops on accessing government assistance programs, filing insurance claims, and navigating the recovery process after a flood.
S. Public service announcements disseminating critical flood warnings, evacuation orders, and safety instructions through radio, television, and online platforms.
T. Emergency preparedness websites offering comprehensive resources, including flood maps, shelter locations, evacuation routes, and contact information for emergency services.
U. Training programs for emergency responders, public officials, and volunteers on incident command systems, emergency operations coordination, and interagency collaboration during flood events.
V. Community forums or town hall meetings providing updates on flood risk assessments, infrastructure projects,

and policy initiatives aimed at improving flood resilience and disaster management at the local level.

W. By tailoring educational materials and resources to the specific needs and preferences of diverse audiences, communities can effectively raise awareness about flood risks, empower individuals and organizations to take proactive measures, and enhance overall preparedness, response, and recovery capabilities in the face of flooding and other natural disasters.

7. Outreach Campaigns

    Implement targeted outreach campaigns to raise awareness, build knowledge, and promote behavior change related to flooding. Use a mix of traditional and digital channels to reach diverse audiences.

8. Training and Workshops

    Offer training sessions and workshops to equip community members with the knowledge and skills needed to prepare for and respond to flooding events. Provide information on evacuation routes, emergency contacts, and preparedness kits.

9. Monitoring and Evaluation

    Continuously monitor and evaluate the effectiveness of education and outreach efforts. Collect feedback from stakeholders and track key metrics such as audience reach, knowledge levels, and behavior change.

10. Adaptation and Improvement
    Based on evaluation findings, adjust and refine education and outreach activities as needed to improve effectiveness and impact over time. Stay flexible and responsive to evolving community needs and challenges related to flooding.

By following these steps, you can effectively migrate from flooding with Community Education and Outreach, building resilience and empowering communities to better prepare for and respond to flooding events.

Green Infrastructure Retrofitting

Plan:

- Retrofit existing infrastructure with green infrastructure features

Green infrastructure refers to natural and engineered systems that mimic the functions of natural ecosystems to manage stormwater and provide other environmental benefits.

Action

1. Reduced Runoff
    Green infrastructure features such as green roofs, permeable pavements, rain gardens, and bioswales absorb and detain rainfall, allowing it to infiltrate into the ground. This reduces the volume of stormwater runoff entering drainage systems and lowers the risk of flooding during heavy rainfall events.

    Reduced runoff through green infrastructure retrofitting can mitigate flooding by managing stormwater at its source, reducing the volume and velocity of runoff, and improving overall drainage and infiltration capabilities. Here's how reduced runoff achieved through green infrastructure retrofitting helps mitigate flooding:

    A. Increased Infiltration
    Green infrastructure features such as permeable pavements, rain gardens, bioswales, and vegetated strips allow rainwater to infiltrate into the ground rather than running off into storm drains. By increasing infiltration rates, green infrastructure reduces the volume of runoff entering drainage systems, thereby mitigating the risk of localized flooding during heavy rainfall events.

B. Stormwater Retention

Green infrastructure elements, such as detention basins, constructed wetlands, and retention ponds, capture and temporarily store stormwater runoff, slowing down its release into downstream water bodies. By retaining excess water and releasing it gradually over time, green infrastructure helps attenuate peak flows, reduce the risk of flash flooding, and protect downstream areas from inundation.

C. Improved Drainage Capacity

Green infrastructure retrofitting can enhance the capacity of existing drainage systems to accommodate stormwater runoff by reducing the volume and intensity of runoff entering the system. By diverting runoff to green spaces and naturalized areas, such as parks, greenways, and floodplain restoration sites, green infrastructure reduces the burden on conventional drainage infrastructure, preventing overloading and reducing the likelihood of system failures during heavy rain events.

D. Floodplain Reconnection

Green infrastructure projects often involve restoring and reconnecting natural floodplains, wetlands, and riparian areas to their historical hydrological functions. By allowing floodwaters to spread out horizontally and infiltrate into the soil, floodplain reconnection reduces the depth and velocity of floodwaters, mitigating the risk of riverine flooding and protecting adjacent communities and infrastructure from inundation.

E. Erosion Control

Green infrastructure practices, such as riparian buffers, streambank stabilization, and erosion control measures, help reduce erosion and sedimentation in water bodies, improving water quality and maintaining natural drainage pathways. By stabilizing stream channels and preventing sediment buildup, green infrastructure reduces the risk of channel blockages, flooding, and downstream impacts caused by sediment-laden runoff during storms.

F. Biodiversity and Habitat Benefits
Green infrastructure projects provide multiple co-benefits, including enhanced biodiversity, improved habitat connectivity, and ecosystem resilience. By creating green corridors and wildlife habitats, green infrastructure promotes natural floodplain processes, such as soil retention, groundwater recharge, and flood attenuation, which contribute to flood risk reduction and ecosystem health.

Reduced runoff achieved through green infrastructure retrofitting plays a vital role in mitigating flooding by managing stormwater at its source, improving drainage capacity, reducing peak flows, and enhancing natural floodplain functions. By incorporating green infrastructure practices into urban and rural landscapes, communities can build resilience to flooding, protect water resources, and create healthier, more sustainable environments for current and future generations.

2. Increased Infiltration
Vegetation and soil in green infrastructure elements help to increase infiltration rates, allowing rainwater to percolate into the ground rather than flowing over impervious surfaces. This helps recharge groundwater and reduces surface runoff, mitigating flood risk.

Increased infiltration through green infrastructure retrofitting helps mitigate flooding by reducing the volume of stormwater runoff that enters drainage systems and overwhelms them during heavy rainfall events. Here's how increased infiltration achieved through green infrastructure retrofitting contributes to flood mitigation:

A. Reduced Peak Flows

When stormwater infiltrates into the ground instead of running off into drainage systems, it slows down the rate at which water reaches watercourses and channels. This reduction in runoff volume and velocity helps to attenuate peak flows, meaning that floodwaters are spread out over a longer period, reducing the risk of flash flooding and decreasing the intensity of flooding downstream.

B. Improved Groundwater Recharge

Green infrastructure features such as permeable pavements, rain gardens, and bioswales allow stormwater to percolate into the soil, replenishing groundwater supplies. Increased groundwater recharge helps to sustain baseflow in rivers and streams during dry periods, reducing the likelihood of low flow conditions and enhancing the overall resilience of aquatic ecosystems to drought and flood events.

C. Sustainable Drainage

Green infrastructure promotes sustainable drainage practices by mimicking natural hydrological processes and reducing reliance on conventional drainage systems. By capturing and infiltrating stormwater onsite, green infrastructure helps to alleviate pressure on stormwater conveyance infrastructure, reducing the risk of system overload and minimizing the need for costly upgrades or expansion of drainage networks.

D. Floodplain Restoration
Green infrastructure projects often involve restoring and reconnecting natural floodplains and wetlands to their historical functions. By promoting infiltration and allowing floodwaters to spread out horizontally across restored floodplain areas, green infrastructure helps to increase the storage capacity of river systems, attenuate flood peaks, and reduce the risk of riverine flooding in adjacent communities.

E. Water Quality Improvement
Increased infiltration through green infrastructure helps to filter pollutants and contaminants from stormwater runoff, improving water quality in rivers, streams, and groundwater aquifers. By reducing the volume of polluted runoff entering water bodies, green infrastructure mitigates the risk of waterborne diseases, protects aquatic habitats, and enhances the overall health of ecosystems downstream.

F. Climate Resilience
Green infrastructure enhances the resilience of communities to climate change impacts, including increased frequency and intensity of rainfall events. By promoting infiltration and reducing surface runoff, green infrastructure helps to manage excess water during storms, reducing the risk of flooding, erosion, and property damage in urban and rural areas vulnerable to climate-related hazards.

Increased infiltration achieved through green infrastructure retrofitting plays a vital role in mitigating flooding by managing stormwater at its source, reducing runoff volumes, and enhancing the capacity of natural and built environments to absorb and retain excess water during heavy rainfall events. By incorporating green infrastructure practices into urban and rural landscapes, communities can build resilience to flooding, protect water resources, and create healthier, more sustainable environments for current and future generations.

3. Stormwater Storage
   Some green infrastructure features, such as retention ponds and wetlands, act as temporary storage reservoirs for stormwater. They capture and retain excess water during heavy rainfall events, gradually releasing it over time to reduce peak flows in downstream waterways and minimize flooding.

   Stormwater storage, as part of green infrastructure retrofitting, can effectively mitigate flooding by capturing and temporarily storing excess rainwater during heavy rainfall events, thereby reducing the volume and intensity of runoff that enters drainage systems. Here's how stormwater storage in green infrastructure retrofitting contributes to flood mitigation:

   Peak Flow Reduction
   A. By storing stormwater runoff, green infrastructure elements such as detention basins, rain barrels, cisterns, and underground storage tanks help reduce the peak flow of water into rivers, streams, and drainage systems during storms. This attenuation of peak flows helps prevent flash flooding and minimizes the risk of inundation downstream.

   B. Temporary Water Storage
   Stormwater storage facilities temporarily hold excess rainwater, allowing it to be slowly released over time. This controlled release mitigates the sudden surge of water into drainage systems, alleviating pressure on infrastructure and

reducing the likelihood of overflows, backups, and localized flooding in urban areas.

C. Flood Risk Reduction
By capturing and storing stormwater at its source, green infrastructure reduces the volume of runoff that reaches rivers, streams, and coastal areas, thereby lowering the risk of riverine and coastal flooding. Stormwater storage facilities help buffer the impacts of heavy rainfall events, protecting communities, infrastructure, and ecosystems from the adverse effects of flooding.

D. Floodplain Management
Green infrastructure projects that incorporate stormwater storage features often include floodplain restoration and enhancement measures. By strategically locating storage facilities within flood-prone areas, such as riparian zones and wetlands, green infrastructure helps expand the capacity of natural floodplains to absorb and retain excess water, reducing flood risk for downstream communities.

E. Water Quality Improvement
Stormwater storage facilities facilitate the settling and filtration of pollutants and sediments from runoff, improving water quality in rivers, lakes, and groundwater aquifers. By capturing pollutants before they reach water bodies, green infrastructure helps protect aquatic habitats, enhance biodiversity, and reduce the risk of contamination-related flooding.

F. Drought Resilience
In addition to flood mitigation, stormwater storage in green infrastructure provides a valuable water resource during periods of drought and water scarcity. Stored rainwater can be

used for landscape irrigation, toilet flushing, and other non-potable uses, reducing reliance on freshwater sources and enhancing community resilience to water shortages.

G. Climate Adaptation

Green infrastructure with stormwater storage capabilities contributes to climate adaptation efforts by managing the impacts of climate change, such as increased frequency and intensity of rainfall events. By capturing and storing excess rainwater, green infrastructure helps communities adapt to changing weather patterns, reduce flood risk, and build resilience to climate-related hazards.

Stormwater storage in green infrastructure retrofitting plays a crucial role in mitigating flooding by capturing, storing, and managing excess rainwater, thereby reducing the volume and intensity of runoff that enters drainage systems and protecting communities and ecosystems from the impacts of flooding. By incorporating stormwater storage features into urban and rural landscapes, communities can enhance flood resilience, improve water quality, and create more sustainable environments for current and future generations.

4. Natural Floodplain Protection

Preserving and restoring natural floodplains through green infrastructure interventions allows for the expansion of flood storage areas. Floodplains can temporarily store floodwaters, reducing the risk of downstream flooding and protecting adjacent properties from inundation.

Floodplain protection, when combined with green infrastructure retrofitting and natural flood management techniques, can effectively mitigate flooding by preserving and enhancing natural floodplain functions, reducing flood risk, and promoting sustainable floodplain management practices. Here's how floodplain protection contributes to flood mitigation in conjunction with green infrastructure retrofitting and natural approaches

A. Natural Storage and Conveyance
Floodplain protection preserves and restores natural floodplain areas, such as wetlands, riparian zones, and floodplain forests, which serve as natural storage and conveyance areas for floodwaters. These natural features absorb excess water during heavy rainfall events, slow down the movement of floodwaters, and reduce the volume of runoff that enters downstream channels, thereby mitigating the risk of riverine flooding.

B. Floodwater Retention
By maintaining open space in floodplains and preventing development in high-risk areas, floodplain protection measures allow floodwaters to spread out horizontally across the landscape, increasing the capacity of rivers and streams to accommodate excess water during flood events. This floodwater retention reduces the intensity and duration of flooding, protects downstream communities and infrastructure, and minimizes property damage and loss.

C. Floodplain Reconnection
Green infrastructure retrofitting and natural flood management initiatives often involve reconnecting rivers and streams with their historic floodplains to restore natural floodplain functions and enhance flood resilience. By removing artificial barriers, such as levees and embankments, and restoring hydrological connectivity between rivers and floodplains, floodplain protection measures enable floodwaters to access natural storage areas and dissipate

energy, reducing the risk of channel overbanking and downstream flooding.

D. Flood Hazard Mitigation
Floodplain protection measures, such as land-use planning regulations, zoning ordinances, and floodplain mapping, help identify and mitigate flood hazards in vulnerable areas. By restricting development in flood-prone zones, elevating structures above flood levels, and implementing green infrastructure practices, floodplain protection reduces exposure to flood risk, protects property and infrastructure, and enhances community resilience to flooding.

E. Ecosystem Services
Preserving and restoring natural floodplain ecosystems through floodplain protection measures provide multiple benefits beyond flood mitigation, including biodiversity conservation, water quality improvement, and recreational opportunities. Healthy floodplain ecosystems provide valuable ecosystem services, such as flood buffering, sediment retention, and habitat provision, which contribute to the overall resilience and sustainability of watersheds and riparian areas.

F. Climate Resilience
Floodplain protection enhances community resilience to climate change impacts, such as increased frequency and intensity of rainfall events and rising sea levels. By safeguarding natural floodplain areas and promoting green infrastructure practices, floodplain protection measures help communities adapt to changing environmental conditions, reduce flood risk, and minimize the social, economic, and environmental impacts of flooding.

Floodplain protection, when integrated with green infrastructure retrofitting and natural flood management approaches, plays a critical role in mitigating flooding by preserving and enhancing natural floodplain functions, reducing flood risk, and promoting sustainable floodplain management practices. By incorporating floodplain protection measures into land-use planning, development regulations, and ecosystem restoration initiatives, communities can build resilience to flooding, protect valuable natural resources, and create more sustainable and resilient environments for current and future generations.

5. Streambank Stabilization

Riparian vegetation and vegetated buffer zones along watercourses help stabilize streambanks and reduce erosion. This helps maintain the integrity of river channels, preventing sedimentation and reducing the risk of bank erosion-induced flooding during high-flow events.

Streambank stabilization is a key component of natural floodplain protection that helps mitigate flooding by preventing erosion, maintaining channel stability, and preserving the integrity of riparian ecosystems. Here's how streambank stabilization contributes to flood mitigation in the context of natural floodplain protection:

A. Erosion Control

Streambank stabilization techniques, such as bioengineering, vegetative buffers, and erosion-resistant materials, help prevent soil erosion along riverbanks and stream channels. By stabilizing streambanks and reducing sedimentation, these techniques maintain channel capacity and prevent sediment buildup downstream, reducing the risk of channel blockages and localized flooding during heavy rainfall events.

B. Channel Stability
Stabilizing streambanks helps maintain the natural alignment and profile of rivers and streams, which is essential for channel stability and flood conveyance capacity. By preventing lateral migration and meandering of stream channels, streambank stabilization measures ensure that floodwaters are efficiently conveyed downstream, minimizing the risk of channel avulsion, bank erosion, and flooding in adjacent areas.

C. Floodplain Connectivity
Streambank stabilization projects often incorporate measures to enhance connectivity between rivers and their floodplains, such as vegetated floodplain benches, floodplain terraces, and meander reconnection. By allowing floodwaters to access natural floodplain areas, these measures increase floodwater storage capacity, reduce flow velocities, and attenuate flood peaks, mitigating the risk of riverine flooding and protecting downstream communities and infrastructure.

D. Vegetative Cover
Planting native vegetation along stabilized streambanks helps stabilize soil, reduce surface runoff, and enhance infiltration rates, thereby reducing the volume and velocity of runoff entering streams and rivers during storms. Healthy riparian vegetation also provides shade, habitat, and food sources for aquatic and terrestrial species, enhancing ecosystem resilience and biodiversity in riparian areas.

E. Sediment Management
Streambank stabilization projects often include measures to manage sediment transport and deposition within river systems, such as sediment traps, vegetated swales, and in-

stream structures. By trapping sediment and preventing excessive deposition, these measures help maintain channel capacity, reduce the risk of sediment-related flooding, and preserve aquatic habitats and water quality downstream.

F. Long-Term Stability

Streambank stabilization techniques focus on promoting natural geomorphic processes and ecological functions to achieve long-term stability and resilience of stream channels and riparian ecosystems. By mimicking natural streambank features and hydrological processes, these techniques create self-sustaining and adaptive systems that can withstand hydrological disturbances, such as floods and droughts, and maintain ecosystem services over time.

G. Community Engagement

Streambank stabilization projects often involve collaboration with local communities, landowners, and stakeholders to ensure the success and sustainability of floodplain protection efforts. Community engagement activities, such as volunteer stewardship programs, educational workshops, and public outreach campaigns, raise awareness about the importance of streambank stabilization for flood mitigation and ecosystem health, fostering a sense of stewardship and collective responsibility for riparian resources.

Streambank stabilization plays a crucial role in natural floodplain protection by preventing erosion, maintaining channel stability, and enhancing the resilience of riparian ecosystems to floods and other hydrological disturbances. By incorporating streambank stabilization measures into floodplain management strategies, communities can reduce flood risk, protect valuable natural resources, and promote sustainable and resilient riparian environments for current and future generations.

6. Improved Water Quality
   Green infrastructure practices filter pollutants and sediments from stormwater runoff, improving water quality in rivers, lakes, and streams. Cleaner watercourses are better able to convey floodwaters and reduce the risk of blockages and backups in drainage systems. See previous chapters

7. Enhanced Urban Resilience
   Integrating green infrastructure into urban planning and design enhances the resilience of communities to climate change and extreme weather events. By diversifying stormwater management approaches and reducing reliance on traditional "gray" infrastructure, cities can better adapt to changing environmental conditions and reduce flood vulnerability. See previous chapters

Green infrastructure offers a sustainable and multifunctional approach to flood mitigation, providing numerous environmental, social, and economic benefits while reducing the risk of flooding and enhancing the resilience of communities to climate-related hazards.

Desalination and Water Recycling

Plan:

- Diversify water supply sources through desalination and water recycling initiatives to reduce reliance on rainfall-dependent sources and ensure water availability during dry periods. Here's how

Action

1. Reduced Vulnerability to Drought

Desalination and water recycling provide alternative sources of freshwater that are not dependent on rainfall. In regions where droughts are common, these technologies offer a reliable and resilient water supply, reducing the vulnerability of communities to water shortages during prolonged dry periods when rainfall-dependent sources may be insufficient.

A. Diversification of Water Sources

Desalination provides an alternative source of freshwater that is not dependent on rainfall or surface water availability. In flood-prone areas, where rainfall can be highly variable and unreliable, desalination offers a consistent and resilient water supply during periods of drought when traditional water sources may be depleted or contaminated due to flooding.

B. Reliable Water Supply

Desalination plants can operate continuously regardless of weather conditions, making them particularly valuable during droughts when surface water sources may be scarce or compromised. By providing a reliable and secure water supply, desalination helps ensure continuity of essential services, such as drinking water supply, agricultural irrigation, and industrial processes, even during extended dry periods when other water sources may be inadequate.

C. Reduction of Pressure on Surface Water Sources

During droughts, surface water sources such as rivers, lakes, and reservoirs are often overexploited as communities compete for limited water resources. Desalination reduces the pressure on these vulnerable water sources by providing an alternative source of freshwater that does not compete with surface water supplies. By diversifying water supply sources, desalination helps alleviate stress on ecosystems and reduces the risk of water shortages and conflicts during droughts.

D. Water Conservation

Desalination promotes water conservation by encouraging the efficient use and reuse of water resources. In drought-prone areas, where water conservation is essential for maintaining water security, desalination facilitates the treatment and reuse of wastewater for non-potable applications such as landscape irrigation, industrial processes, and toilet flushing. By maximizing the use of available water resources, desalination helps minimize water waste and reduce overall water demand, enhancing drought resilience in flood-prone areas.

E. Long-Term Planning

Investing in desalination infrastructure enables communities to adopt a long-term planning approach to water resource management that considers future climate projections and water availability scenarios. By diversifying water supply sources and reducing reliance on rainfall-dependent sources, desalination helps communities adapt to changing climatic conditions, mitigate the impacts of droughts and floods, and ensure water security for current and future generations.

Desalination plays a crucial role in reducing vulnerability to drought in flood-prone areas by providing a reliable, resilient, and drought-resistant water supply that complements traditional water sources. By diversifying water supply sources, promoting water conservation, and facilitating long-term planning, desalination helps enhance drought resilience, mitigate water scarcity, and ensure sustainable water management practices in flood-prone regions.

2. Buffer Against Flooding

In areas prone to flooding, traditional water sources such as rivers and reservoirs can become contaminated or compromised during flood events. Desalination plants and water recycling facilities, often located inland and away from flood-prone areas, provide a reliable and secure water supply that is less susceptible to contamination or disruption during floods, ensuring continuity of water supply for essential services and emergency response efforts.

To buffer against flooding in flood-prone areas, several strategies can be implemented:

A.Floodplain Preservation

Preserving natural floodplains helps absorb excess water during flood events. These areas act as natural sponges, allowing water to spread out horizontally, slowing its flow, and reducing the intensity of flooding downstream.

B.Riparian Vegetation

Planting and maintaining vegetation along riverbanks and streams can stabilize soil, reduce erosion, and absorb water during flood events. Healthy riparian zones act as buffers, slowing down floodwaters and reducing the risk of bank erosion and channel scouring.

C. Green Infrastructure

Implementing green infrastructure practices such as rain gardens, vegetated swales, and permeable pavements can help absorb and infiltrate stormwater runoff, reducing the volume and velocity of water entering watercourses during heavy rainfall events. Green infrastructure features act as natural buffers, mitigating the impacts of flooding and reducing the risk of urban drainage system overload.

D. Floodplain Restoration

Restoring degraded or modified floodplain areas to their natural state can enhance flood buffering capacity by allowing floodwaters to access natural storage areas, such as wetlands, ponds, and floodplain depressions. Floodplain restoration projects can help attenuate flood peaks, reduce flood risk, and protect downstream communities and infrastructure.

E. Floodplain Zoning and Land Use Planning

Implementing land use regulations and zoning ordinances that restrict development in flood-prone areas helps prevent encroachment into high-risk zones and reduces exposure to flood hazards. By preserving natural floodplains and limiting development in flood-prone areas, communities can minimize property damage, protect critical infrastructure, and enhance overall flood resilience.

F. Flood Control Structures

Constructing flood control structures such as levees, floodwalls, and retention basins can help mitigate flooding by containing and diverting floodwaters away from populated areas. These structures act as physical barriers, providing protection against inundation and reducing the extent and severity of flooding in flood-prone areas.

G. Community Education and Preparedness

Educating residents about flood risks, emergency preparedness, and evacuation procedures is essential for enhancing community resilience to flooding. By raising awareness and promoting proactive measures such as floodproofing, emergency kits, and evacuation plans, communities can minimize the impacts of flooding and ensure the safety and well-being of residents during flood events.

Buffering against flooding in flood-prone areas requires a combination of natural and engineered solutions, land use planning measures, and community engagement efforts. By implementing these strategies in a comprehensive and integrated manner, communities can reduce vulnerability to flooding, protect lives and property, and promote sustainable flood risk management practices in flood-prone regions.

3. Improved Water Quality

Desalination and water recycling technologies produce high-quality treated water that meets stringent standards for drinking water, industrial processes, and irrigation. By diversifying water supply sources, communities can reduce their reliance on surface water sources that may be prone to pollution from runoff, sedimentation, and other contaminants associated with flooding, thereby improving overall water quality and public health outcomes.

4. Water Conservation

Water recycling initiatives promote the efficient use and conservation of water resources by capturing, treating, and reusing wastewater for non-potable applications such as landscape irrigation, industrial processes, and toilet flushing. By reducing demand for freshwater from traditional sources, water recycling helps alleviate pressure on water supplies and ecosystems, particularly in water-stressed regions where competing demands for water are high.

5. Climate Resilience

Desalination and water recycling contribute to climate resilience by providing alternative water sources that are less vulnerable to climate variability and extreme weather events, such as floods and droughts. By diversifying water supply sources and reducing reliance on rainfall-dependent sources, communities can adapt to changing climate conditions, mitigate the impacts of water scarcity and flooding, and enhance their overall resilience to climate-related hazards.

6. Sustainable Water Management

Desalination and water recycling initiatives support sustainable water management practices by promoting the efficient use, treatment, and reuse of water resources. These technologies help minimize the environmental footprint of water supply systems, reduce energy consumption, and protect ecosystems by minimizing the extraction of freshwater from natural sources and reducing the discharge of untreated wastewater into water bodies.

7. Long-Term Planning

By investing in desalination and water recycling infrastructure, communities can adopt a long-term planning approach to water resource management that considers future population growth, urban development, and climate change projections. Diversifying water supply sources provides a strategic hedge against uncertain hydrological conditions and ensures water security for current and future generations, regardless of variations in rainfall patterns or flood risks.

Diversifying water supply sources through desalination and water recycling initiatives offers numerous benefits for flood-prone areas, including enhanced resilience to droughts and floods, improved water quality, increased water conservation, and sustainable water management practices. By integrating these technologies into comprehensive water resource management strategies, communities can enhance their resilience to climate change, reduce vulnerability to water-related hazards, and ensure reliable access to safe and sustainable water supplies for all residents.

Infrastructure Resilience Planning

Plan:

- Incorporate climate change adaptation measures. Mitigating flooding with urban resilience involves a multifaceted approach that integrates various strategies and measures to minimize the impact of flooding on urban areas. In urban areas, subsurface drainage systems can be integrated with stormwater management infrastructure to enhance resilience to flooding. They can be used in combination with green infrastructure features such as permeable pavements, rain gardens, and bioswales to manage stormwater runoff effectively and reduce flood risk. Infrastructure resilience planning involves developing strategies and measures to ensure that critical infrastructure systems can withstand and recover from various hazards, including floods. Conducting a comprehensive risk assessment to identify flood-prone areas, vulnerable infrastructure, and at-risk communities is a crucial step in infrastructure resilience planning.

Action

1. Assessment and Planning

- Conduct a comprehensive risk assessment to identify flood-prone areas, vulnerable infrastructure, and at-risk communities.

- Develop an urban resilience plan that integrates flood mitigation strategies with broader resilience goals, considering factors such as land use planning, infrastructure design, and community engagement.

  A. Identify Critical Infrastructure

Begin by identifying the critical infrastructure in the area under assessment. This includes essential facilities such as hospitals, emergency services, water treatment plants, power plants, transportation networks (roads, bridges, railways), and communication systems. These are the backbone of the community and need to be resilient to flooding.

B. Assess Vulnerability

Evaluate the vulnerability of critical infrastructure to flooding. This involves understanding the susceptibility of infrastructure to flood damage based on factors such as location, elevation, design standards, age, and condition. Vulnerability assessments can be conducted using various tools and techniques, including GIS (Geographic Information Systems) mapping, hydraulic modeling, and historical flood data analysis.

C. Identify Flood-Prone Areas

Use historical flood data, floodplain maps, and hydrological modeling to identify areas prone to flooding. This step helps in understanding the extent and severity of flood risks within the community.

D. Evaluate Socio-Economic Factors

Consider socio-economic factors such as population density, demographics, income levels, and social vulnerabilities. Certain communities, such as low-income neighborhoods or marginalized groups, may be more vulnerable to the impacts of flooding due to limited resources or inadequate infrastructure.

E. Engage Stakeholders

Involve relevant stakeholders, including local government agencies, emergency responders, community organizations, and residents, in the risk assessment process. Stakeholder engagement ensures that diverse perspectives are considered and helps in developing effective resilience strategies that are tailored to the needs of the community.

F. Prioritize Mitigation Measures

Based on the findings of the risk assessment, prioritize mitigation measures to enhance the resilience of infrastructure and protect at-risk communities. This may include upgrading infrastructure to withstand flood events, implementing green infrastructure solutions (e.g., green roofs, permeable pavements), improving early warning systems, and enhancing emergency preparedness and response plans.

G. Integrate Resilience into Planning and Development

Incorporate resilience considerations into land use planning, zoning regulations, building codes, and infrastructure development projects. By integrating resilience into decision-making processes, future development can be designed to reduce flood risk and enhance overall community resilience.

H. Monitor and Review

Continuously monitor and review the effectiveness of resilience measures and update risk assessments as needed. Regular evaluations help in identifying emerging threats, evaluating the performance of existing infrastructure systems, and refining resilience strategies over time.

Through integrating infrastructure resilience planning with comprehensive risk assessment for flood-prone areas, communities can enhance their ability to withstand and recover from flooding events, ultimately reducing the social, economic, and environmental impacts of floods.

2. Green Infrastructure

Green infrastructure can play a crucial role in resilience planning for water management and preventing flooding.

- Implement green infrastructure solutions such as permeable pavements, green roofs, rain gardens, and urban wetlands to absorb and manage stormwater runoff, reducing the risk of flooding.

- Preserve and restore natural floodplains and wetlands to provide natural flood storage and buffer areas.

A. Permeable Pavements

Implement permeable pavements in parking lots, sidewalks, and roads. These surfaces allow rainwater to infiltrate into the ground, reducing surface runoff and lessening the burden on stormwater drainage systems. By promoting infiltration, permeable surfaces help prevent flooding by reducing the volume of water flowing into waterways during heavy rainfall.

B. Green Roofs

Install green roofs on buildings, particularly in urban areas with limited green space. Green roofs are covered with vegetation and soil, which absorb rainwater and reduce runoff. By retaining water, green roofs can delay and reduce the volume of stormwater entering drainage systems, mitigating the risk of localized flooding and reducing pressure on urban infrastructure.

C. Rain Gardens and Bioswales

Construct rain gardens and bioswales in strategic locations, such as along streets or in parking lots, to capture and absorb stormwater. These vegetated features help slow down and filter runoff, allowing water to percolate into the soil and recharge groundwater. By intercepting and managing runoff, rain gardens and bioswales can prevent flooding by reducing the flow of water into drainage systems.

D. Constructed Wetlands

Develop constructed wetlands in flood-prone areas to provide additional storage capacity for stormwater and reduce downstream flooding. Wetlands act as natural sponges, absorbing excess water.

2. Grey Infrastructure Upgrades
   Grey infrastructure upgrades can play a critical role in resilience planning for water management and preventing flooding. There are several ways in which grey infrastructure upgrades can be utilized for flood prevention.

- Upgrade and maintain existing drainage systems, culverts, and stormwater management infrastructure to improve their capacity and efficiency.

- Install flood barriers, levees, and flood walls in high-risk areas to protect critical infrastructure and communities from floodwaters.

A. Stormwater Management Systems

Implementing or upgrading stormwater management systems, including drainage networks, sewers, and detention basins, can significantly reduce the risk of flooding. These systems are designed to convey, store, and control stormwater runoff, preventing it from overwhelming urban areas and causing flooding. Upgrading these systems to increase capacity, improve efficiency, and incorporate innovative technologies can enhance flood resilience.

B. Flood Control Structures

Constructing or enhancing flood control structures such as levees, floodwalls, and dams can provide effective protection against flooding in flood-prone areas. These structures are designed to contain floodwaters within specified boundaries, preventing them from inundating surrounding communities. Upgrading existing flood control infrastructure and implementing new projects in strategic locations can mitigate flood risks and protect vulnerable areas.

C. Pumping Stations and Floodgates

Installing or upgrading pumping stations and floodgates can help manage water levels during flood events, particularly in low-lying or coastal areas. These infrastructure upgrades can facilitate the controlled removal of excess water from urban areas and prevent inundation of critical infrastructure such as roads, utilities, and buildings. Investing in resilient pumping and gate systems can improve flood response capabilities and minimize flood damages.

D. Channelization and Dredging

Channelizing waterways and dredging channels can improve the conveyance capacity of rivers, streams, and drainage channels, reducing the risk of localized flooding. These measures involve reshaping or deepening watercourses to enhance flow capacity and minimize obstructions. Upgrading channel infrastructure through maintenance dredging, vegetation management, and erosion control measures can optimize hydraulic performance and mitigate flood hazards.

E. Retention and Detention Facilities

Constructing retention and detention facilities, such as ponds, reservoirs, and wetlands, can help attenuate peak flows and reduce downstream flooding. These facilities temporarily store excess stormwater and gradually release it at controlled rates, thereby mitigating the impacts of heavy rainfall events. Upgrading existing facilities and implementing new ones in strategic locations can enhance flood resilience and protect downstream communities.

F. Integrated Floodplain Management

Implementing integrated floodplain management strategies, including land use planning, zoning regulations, and floodplain mapping, can help minimize flood risks and protect vulnerable areas. By restricting development in flood-prone zones, preserving natural floodplains, and promoting resilient building practices, communities can reduce exposure to flood hazards and enhance overall resilience to flooding events.

G. Emergency Response and Early Warning Systems

Enhancing emergency response capabilities and early warning systems can improve flood preparedness and reduce the impacts of flooding. Investing in flood monitoring technology, communication networks, and evacuation plans can enable timely and effective responses to flood events, saving lives and reducing property damage. Upgrading emergency infrastructure and coordination mechanisms can strengthen community resilience and ensure a coordinated flood response.

H. Infrastructure Maintenance and Rehabilitation

Regular maintenance and rehabilitation of existing grey infrastructure assets are essential for ensuring their effectiveness in flood prevention. Investing in infrastructure maintenance programs, such as cleaning debris from drainage systems, repairing deteriorating structures, and upgrading aging infrastructure components, can prolong the service life of assets and minimize flood risks. Prioritizing infrastructure resilience and investing in proactive maintenance measures can safeguard communities against flooding and enhance long-term resilience.

Leveraging grey infrastructure upgrades in resilience planning for water management, communities can effectively prevent flooding, reduce vulnerability to flood hazards, and enhance overall resilience to climate-related risks. Integrating grey infrastructure solutions with nature-based approaches and community engagement initiatives can create comprehensive flood resilience strategies that protect lives, property, and critical infrastructure.

4. Community Engagement and Education

Community engagement and education are essential components of resilience planning for water management and preventing flooding.

- Engage with local communities to raise awareness about flood risks, preparedness, and resilience measures.

- Provide training and resources to help residents and businesses develop flood emergency plans and implement mitigation measures.

A. Raise Awareness

Educate community members about flood risks, including the causes and potential impacts of flooding in their area. Use various communication channels such as community meetings, workshops, educational materials, social media, and local media outlets to disseminate information about flood hazards, emergency preparedness, and resilience strategies.

B. Involve Stakeholders

Engage stakeholders, including residents, businesses, local government officials, emergency responders, community organizations, and other relevant groups, in the resilience planning process. Encourage active participation and input from diverse perspectives to ensure that resilience strategies are inclusive, culturally appropriate, and address the needs and concerns of all community members.

C. Identify Local Knowledge

Tap into local knowledge and expertise by involving community members in data collection, risk assessment, and decision-making processes. Local residents often have valuable insights into flood vulnerabilities, historical flood events, and informal coping strategies that can inform resilience planning efforts and enhance the effectiveness of flood prevention measures.

D. Promote Risk Reduction Practices

Empower community members to take proactive measures to reduce their vulnerability to flooding. Provide information and resources on flood-resistant building techniques, property-level flood protection measures, flood insurance options, evacuation planning, and emergency response procedures. Encourage individuals and businesses to implement resilience measures that can mitigate flood risks and enhance community preparedness.

E. Foster Collaboration

Facilitate collaboration and partnerships among stakeholders to leverage collective expertise, resources, and networks for resilience planning and implementation. Encourage collaborative initiatives between government agencies, non-profit organizations, academic institutions, private sector entities, and community groups to address complex challenges related to flood prevention and resilience building.

F. Support Community-led Initiatives

Empower communities to develop and implement their own resilience projects and initiatives. Provide technical assistance, funding opportunities, and capacity-building support to grassroots organizations, neighborhood associations, and volunteer groups interested in undertaking resilience projects such as green infrastructure installations, floodplain restoration efforts, or community-based monitoring programs.

G. Build Social Capital

Strengthen social connections, trust, and cohesion within communities to enhance collective resilience. Foster a sense of belonging and solidarity among residents through social events, neighborhood gatherings, and collaborative activities. Building social capital can enhance community resilience by facilitating communication, cooperation, and mutual support during times of crisis.

F. Promote Education and Training

Offer educational programs, training workshops, and skill-building activities to empower community members with the knowledge, skills, and tools needed to respond effectively to flooding and other hazards. Provide training on topics such as flood risk assessment, emergency preparedness, disaster response, first aid, and psychological resilience to enhance community resilience and preparedness.

Actively engaging and educating community members in resilience planning for water management, communities can enhance their capacity to prevent flooding, reduce vulnerability to flood hazards, and build more resilient and sustainable communities. Community engagement fosters ownership, collaboration, and innovation, leading to more effective and inclusive resilience strategies that benefit everyone.

5. Integrated Water Management

Integrated Water Management (IWM) is a holistic approach that considers the entire water cycle, from rainfall to runoff, infiltration, storage, treatment, and reuse. Integrating IWM principles into resilience planning for water can be highly effective in preventing flooding.

- Adopt integrated water management approaches that consider the entire water cycle, including stormwater management, wastewater treatment, and water supply.

- Coordinate with neighboring jurisdictions and stakeholders to implement regional solutions and address upstream/downstream impacts.

A. Stormwater Capture and Storage

Implement strategies to capture and store stormwater runoff for reuse or infiltration. This can include constructing rainwater harvesting systems, retention ponds, and underground storage tanks to capture excess rainfall and prevent it from contributing to flooding downstream.

B. Green Infrastructure

Integrate green infrastructure elements such as permeable pavements, bioswales, rain gardens, and green roofs into urban landscapes. These natural or nature-based solutions help absorb and slow down stormwater runoff, reducing the volume of water entering drainage systems and mitigating the risk of flooding.

C. Floodplain Management

Implement measures to manage floodplains more effectively, such as restoring natural floodplain areas and establishing setback requirements for development in flood-prone zones. By preserving or restoring floodplains, communities can provide additional storage capacity for floodwaters and reduce the risk of downstream flooding.

D. Flood Control Structures

Construct or upgrade flood control structures, such as levees, floodwalls, and detention basins, to manage flood risk in vulnerable areas. These engineered solutions can help contain floodwaters within designated boundaries and protect communities from inundation during extreme weather events.

E. Stormwater Treatment and Pollution Control

Integrate stormwater treatment measures into water management systems to improve water quality and reduce pollution. Implementing features such as constructed wetlands, biofiltration systems, and sedimentation basins can help remove pollutants from stormwater runoff before it enters water bodies, enhancing both flood resilience and environmental sustainability.

F. Smart Water Infrastructure

Deploy smart water management technologies and systems to monitor, analyze, and optimize water resources in real-time. Utilize sensors, data analytics, and predictive modeling to anticipate flood events, optimize water flow, and manage water infrastructure more efficiently, reducing the impact of flooding on communities.

G. Community Engagement and Education

Engage stakeholders, including residents, businesses, and government agencies, in the planning and implementation of integrated water management strategies. Raise awareness about the importance of water conservation, flood prevention, and sustainable water management practices through educational campaigns, workshops, and outreach initiatives.

H. Multi-Stakeholder Collaboration

Foster collaboration and coordination among various stakeholders involved in water management, including government agencies, utilities, developers, and community organizations. Establish partnerships to pool resources, share expertise, and develop integrated solutions that address both water quantity and quality issues while enhancing flood resilience.

Integrating Integrated Water Management principles into resilience planning for water, communities can effectively prevent flooding, reduce vulnerability to extreme weather events, and build more resilient and sustainable water systems. This holistic approach considers the interconnectedness of water resources and ecosystems, leading to more effective and adaptive solutions that benefit both people and the environment.

6. Risk Reduction Measures

Risk reduction measures play a crucial role in resilience planning for water management and preventing flooding.

- Implement land use planning policies that discourage development in flood-prone areas and promote resilient building design and construction standards.

- Encourage the adoption of flood insurance and incentivize property owners to invest in flood-resistant measures such as elevating structures and retrofitting buildings.

A. Flood Risk Assessment

Conduct comprehensive flood risk assessments to identify areas prone to flooding and understand the potential impacts of flood events on communities, infrastructure, and the environment. Utilize data, modeling, and mapping techniques to assess flood hazards, vulnerabilities, and risks, enabling informed decision-making and prioritization of mitigation efforts.

B. Land Use Planning and Zoning Regulations

Integrate flood risk considerations into land use planning and zoning regulations to guide development away from flood-prone areas and minimize exposure to flood hazards. Establish setback requirements, building elevation standards, and floodplain management regulations to reduce the risk of property damage and protect public safety.

C. Floodplain Management

Implement measures to manage and preserve natural floodplains, such as restoring wetlands, establishing riparian buffers, and limiting development in flood-prone zones. Protecting floodplain areas helps absorb and attenuate floodwaters, reducing the severity of flooding downstream and protecting communities from inundation.

D. Infrastructure Resilience Upgrades

Upgrade critical infrastructure systems to enhance their resilience to flooding and minimize disruption during extreme weather events. Retrofitting buildings, roads, bridges, and utilities with flood-resistant design features can reduce the risk of damage and ensure continuity of essential services during floods, improving community resilience and recovery.

E. Stormwater Management Systems

Implement stormwater management systems to capture, store, and treat stormwater runoff, reducing the volume and velocity of water entering drainage systems and waterways. Constructing retention ponds, bioswales, permeable pavements, and green infrastructure features helps mitigate the risk of urban flooding and improve water quality while enhancing community resilience.

F. Early Warning Systems and Emergency Preparedness

Develop and implement early warning systems to provide timely alerts and evacuation notifications to residents in flood-prone areas. Establish emergency preparedness plans, evacuation routes, and shelters to ensure a coordinated response to flood events and minimize the impact on lives and property.

G. Community Education and Outreach

Educate community members about flood risks, emergency preparedness, and resilience measures through outreach campaigns, workshops, and educational materials. Promote flood insurance enrollment, property protection measures, and sustainable water management practices to empower individuals and communities to mitigate flood risks and build resilience.

H. Multi-Stakeholder Collaboration

Foster collaboration and coordination among government agencies, utilities, emergency responders, community organizations, and other stakeholders involved in water management and flood resilience efforts. Establish partnerships to share resources, expertise, and best practices, enabling a holistic and integrated approach to flood risk reduction and resilience planning.

Implementing risk reduction measures in resilience planning for water, communities can effectively prevent flooding, reduce vulnerability to extreme weather events, and build more resilient and sustainable water systems. These measures help protect lives, property, and critical infrastructure while promoting long-term resilience and adaptation to climate change impacts.

7. Early Warning Systems

Early warning systems (EWS) are critical components of resilience planning for water management and preventing flooding.

- Establish robust early warning systems that utilize real-time monitoring, predictive modeling, and communication technologies to alert residents and authorities about impending flood events.

- Ensure that emergency response plans are in place and regularly tested to effectively coordinate evacuation, rescue, and relief efforts.

A. Flood Forecasting and Monitoring

Implement advanced hydrological modeling and monitoring systems to predict flood events and monitor water levels in rivers, streams, and drainage systems. Utilize real-time data from gauges, sensors, and satellite imagery to assess precipitation, river flows, and soil moisture conditions, enabling early detection of flood risks and timely warnings to communities.

B. Alert and Notification Systems

Establish robust alert and notification systems to disseminate timely warnings to residents in flood-prone areas. Utilize multiple communication channels, including text messages, sirens, social media, and mobile apps, to reach diverse populations and ensure broad coverage. Develop clear and concise warning messages that convey the severity of the flood threat and provide actionable guidance on evacuation routes, shelter locations, and emergency contacts.

C. Community Engagement and Preparedness

Engage communities in flood preparedness and response efforts by raising awareness about the importance of early warning systems and encouraging residents to sign up for alerts. Conduct outreach activities, workshops, and training sessions to educate community members about flood risks, emergency procedures, and resilience measures. Empower individuals and households to develop personal preparedness plans and take proactive steps to protect themselves and their property from flooding.

D. Integration with Emergency Management

Integrate early warning systems with broader emergency management frameworks to ensure a coordinated and effective response to flood events. Establish protocols for activating EWS, coordinating evacuation efforts, and mobilizing emergency responders. Collaborate with local government agencies, first responders, utilities, and community organizations to streamline communication, share information, and coordinate response activities during flood emergencies.

E. Continuous Improvement and Evaluation

Continuously evaluate and refine early warning systems based on feedback from stakeholders, lessons learned from past flood events, and advances in technology and modeling capabilities. Conduct regular drills, exercises, and simulations to test the effectiveness of warning systems and enhance community preparedness. Invest in research and innovation to improve the accuracy, reliability, and timeliness of flood forecasts and warnings, enabling more proactive and targeted risk mitigation efforts.

F.Inclusion of Vulnerable Populations

Ensure that early warning systems are accessible and inclusive for all community members, including vulnerable populations such as elderly individuals, people with disabilities, and non-English speakers. Provide tailored communication strategies, alternative formats, and assistance services to reach marginalized groups and address their specific needs during flood emergencies. Collaborate with community-based organizations and social service agencies to ensure that vulnerable populations are adequately supported and informed during flood events.

G.Interagency Collaboration and Data Sharing

Foster collaboration and data sharing among government agencies, research institutions, and international organizations involved in water management and disaster risk reduction. Exchange data, expertise, and best practices to enhance the effectiveness of early warning systems and improve flood resilience at regional and global scales. Promote cross-border cooperation and information exchange to address transboundary flood risks and support coordinated response efforts in shared river basins and coastal areas.

Leveraging early warning systems in resilience planning for water, communities can enhance their capacity to prevent flooding, reduce vulnerability to extreme weather events, and protect lives, property, and critical infrastructure. Early detection of flood risks and timely dissemination of warnings enable communities to take proactive measures to mitigate impacts, evacuate safely, and recover more quickly from flood emergencies.

8. Adaptive Management and Learning

Implementing these strategies in a coordinated and integrated manner, urban areas can enhance their resilience to flooding and minimize the social, economic, and environmental impacts of extreme weather events. Adaptive management and learning are essential concepts in resilience planning for water management and preventing flooding.

- Continuously monitor and evaluate the effectiveness of flood mitigation measures, and adapt strategies based on lessons learned and changing conditions.

- Facilitate a culture of resilience by promoting innovation, collaboration, and knowledge sharing among stakeholders.

A. Continuous Monitoring and Assessment

Implement a system for continuous monitoring and assessment of flood risks, hydrological conditions, and resilience measures. Utilize data from weather stations, stream gauges, remote sensing, and community feedback to assess the effectiveness of flood prevention measures and identify emerging risks or vulnerabilities.

B. Iterative Planning and Decision-Making

Adopt an iterative approach to resilience planning, where decisions are made based on the best available information and adjusted over time in response to changing conditions. Use adaptive management frameworks, such as the Plan-Do-Check-Act cycle, to evaluate the performance of flood prevention measures, learn from experience, and make informed adjustments to resilience strategies as needed.

C. Scenario Planning and Modeling

Develop scenarios and predictive models to explore potential future flood scenarios, considering factors such as climate change, land use changes, and population growth. Use scenario planning exercises to identify plausible futures, assess the resilience of existing infrastructure and policies, and explore alternative strategies for mitigating flood risks under different scenarios.

D. Stakeholder Engagement and Collaboration

Engage stakeholders, including government agencies, local communities, scientists, and decision-makers, in collaborative learning processes. Facilitate dialogue, knowledge sharing, and joint problem-solving to build consensus around resilience goals, identify shared priorities, and co-develop adaptive management strategies that integrate diverse perspectives and expertise.

E. Experimentation and Innovation

Encourage experimentation and innovation in resilience planning by piloting new approaches, technologies, and policies in real-world settings. Establish demonstration projects, living laboratories, or adaptive management experiments to test the feasibility, effectiveness, and scalability of innovative flood prevention measures, such as green infrastructure, ecosystem-based approaches, or nature-based solutions.

F. Capacity Building and Training

Invest in capacity building and training programs to enhance the knowledge, skills, and adaptive capacity of stakeholders involved in resilience planning and implementation. Provide training on topics such as flood risk assessment, climate adaptation, ecosystem services, and participatory decision-making to empower communities and decision-makers to respond effectively to changing flood risks.

G. Learning Networks and Communities of Practice

Foster learning networks and communities of practice to facilitate knowledge exchange, peer-to-peer learning, and collaboration among practitioners, researchers, and policymakers working on flood resilience issues. Create platforms for sharing best practices, case studies, lessons learned, and success stories to inspire innovation and replication of successful approaches in other contexts.

F. Feedback Loops and Evaluation

Establish feedback loops and mechanisms for evaluating the effectiveness of resilience measures, soliciting input from stakeholders, and incorporating lessons learned into future planning efforts. Conduct regular evaluations, post-project reviews, and after-action assessments to assess the outcomes, impacts, and unintended consequences of resilience interventions, and use this information to inform adaptive management decisions and improve future planning efforts.

Integrating adaptive management and learning principles into resilience planning for water management, communities can enhance their capacity to prevent flooding, reduce vulnerability to extreme weather events, and build more resilient and sustainable water systems. Adaptive management approaches enable communities to respond flexibly to changing conditions, learn from experience, and continuously improve their resilience over time, leading to more effective and adaptive flood prevention strategies.

Collaboration and Partnerships

- Plan:

- Facilitate collaboration among government agencies, private sector stakeholders, academic institutions, and communities to develop and implement integrated flood management strategies that address the complex challenges posed by heavy rainfall and urbanization.

Action

Facilitating collaboration among diverse stakeholders is crucial for developing and implementing integrated flood management strategies, especially regarding permeable pavements.

1. Establish Multi-Stakeholder Platforms

Create multi-stakeholder platforms or working groups comprising representatives from government agencies, private sector stakeholders (such as developers, construction companies, and engineering firms), academic institutions, community organizations, and relevant experts. These platforms serve as forums for dialogue, knowledge sharing, and collaborative decision-making on flood management strategies, including permeable pavements.

2. Convene Stakeholder Workshops and Forums

Organize stakeholder workshops, forums, or roundtable discussions to foster collaboration and consensus-building around permeable pavement initiatives. Provide opportunities for stakeholders to exchange ideas, share experiences, and discuss challenges and opportunities related to the design, implementation, and maintenance of permeable pavement systems for flood management.

3. Promote Knowledge Exchange and Capacity Building

Facilitate knowledge exchange and capacity-building activities to enhance stakeholders' understanding of permeable pavement technologies and their potential benefits for flood prevention. Offer training sessions, technical seminars, and educational materials on permeable pavement design, construction techniques, performance monitoring, and maintenance best practices to empower stakeholders with the necessary skills and expertise.

4. Engage Academic and Research Institutions

Collaborate with academic and research institutions to leverage their expertise, resources, and research findings in support of permeable pavement initiatives for flood management. Partner with universities, research centers, and technical institutes to conduct applied research, pilot projects, and demonstration sites that evaluate the effectiveness and performance of permeable pavement systems under different climatic and hydrological conditions.

5. Leverage Public-Private Partnerships

Foster public-private partnerships (PPPs) to leverage the strengths and resources of both sectors in implementing permeable pavement projects for flood management. Engage private sector stakeholders, such as developers, contractors, and material suppliers, in joint ventures, consortia, or collaborative agreements to finance, design, construct, and maintain permeable pavement infrastructure.

6. Develop Collaborative Funding Mechanisms

Establish collaborative funding mechanisms, such as revolving funds, grant programs, or public-private financing schemes, to support permeable pavement initiatives for flood management. Pool financial resources from government agencies, private sector partners, philanthropic organizations, and community contributions to invest in permeable pavement projects that deliver multiple benefits, including flood risk reduction, water quality improvement, and urban greening.

7. Facilitate Demonstration Projects and Pilot Studies

Facilitate the implementation of demonstration projects and pilot studies to showcase the effectiveness and feasibility of permeable pavement solutions for flood management. Collaborate with stakeholders to identify suitable sites, design appropriate demonstration projects, and monitor performance outcomes to inform future decision-making and replication efforts.

8. Promote Policy Alignment and Coordination

Advocate for policy alignment and coordination among government agencies at the local, regional, and national levels to support the mainstreaming of permeable pavement technologies into flood management strategies. Work with policymakers to develop supportive regulatory frameworks, zoning ordinances, building codes, and incentive programs that incentivize the adoption of permeable pavement practices and remove barriers to implementation.

Facilitating collaboration among government agencies, private sector stakeholders, academic institutions, and communities, integrated flood management strategies can be developed and implemented effectively, with permeable pavements playing a key role in preventing flooding and enhancing urban resilience.

In summary, flood management is critical priorities in the 21st century to address the growing challenges of urbanization, climate change, infrastructure vulnerability, economic impacts, public health risks, environmental degradation, and community resilience. Investing in effective flood management strategies is essential to protect lives, property, infrastructure, and the environment, ensuring the safety, prosperity, and sustainability of communities worldwide.

www.ingramcontent.com/pod-product-compliance
Lightning Source LLC
Chambersburg PA
CBHW062107220526
45471CB00010B/3637